Your Health
As The Climate Changes

Your Health As The Climate Changes

Disease and Death on a Warming Planet

DR SCOTT FRASER

**Published in 2025
by Eyewear Publishing Limited
The Black Spring Press Group
London, United Kingdom**

Cover design by Matt Broughton
Typeset by Subash Raghu

All rights reserved
© 2025 Dr Scott Fraser

ISBN 978-1-915406-78-1

The right of Dr Scott Fraser to be identified as author of this work has been asserted in accordance with section 77 of the Copyright, Designs and Patents Act 1988.

www.blackspringpressgroup.com

This Great Earth will persist
Without us latecomers in its midst
From tiny things to armoured beast
We will be the empty chair at the feast.

This book is dedicated to all our children
and grandchildren and…

Contents

Introduction	xvii
Chapter 1 Is the Earth getting warmer (and why)?	1
The boring numbers bit	2
Thermometer Readings	3
Indirect Temperature Measures	4
Evidence from other observations	5
Lived Experience	6
Part of a Natural Cycle?	7
The boring science bit	9
Why is the Earth getting warmer?	9
Carbon Dioxide Numbers	10
How do we know that the rise in CO_2 is caused by humans?	12
Feedback Cycles	13

Chapter 2	**What happens to Nature when the temperature rises (and why that's important to humans)**	**17**
	The Weather	**18**
	Heat and Heatwaves	19
	Droughts	20
	Wildfires	22
	Floods	23
	The Oceans	**25**
	Sea Levels	25
	Ocean Acidification	26
	Warmer Seas	27
	Hurricanes and Tornadoes	**30**
	Plants and Agriculture	**31**
	Animals (except humans)	**34**
	Insects	**36**
	Microorganisms	**38**
	Volcanoes and Earthquakes	**40**
	Ecosystems	**41**
Chapter 3	**What happens to humans as the temperature rises?**	**43**
	How do we normally regulate our body temperature?	43
	How do we avoid getting too hot?	44
	Wet-Bulb Temperature	48

CONTENTS

When the heat gets too high	**49**
Heatwaves	49
The Numbers	50
Hotter and Hotter	52
Chapter 4 Threat by Threat	**56**
Health threats on a warming Planet	**57**
Direct effects to human health	**57**
Earth	58
Wind	60
Fire	63
Water	67
Coastal Areas	75
Snow and Ice melt	77
Indirect threats to human health	**78**
Starvation, malnutrition and food security	79
Insects	84
Microorganisms	85
Changes in larger animal behaviour	89
Air Quality	91
Economics, Health and Climate Change	93
Politics and Conflict and Migration	96
Human Migration	99
Crime and Climate Change	101
Other heat induced behavioural changes	103

Overwhelming of healthcare services	105
Healthcare costs and Climate Change	108
Loss of habitat and Drug discovery	110
Healthcare and its contribution to Climate Change	110

Chapter 5 Disease by Disease 113

Disease specific threats of a warming planet 113

Climate Change and Skin Disease	115
Climate Change and Lung Disease	118
Climate Change and Heart and Circulatory Disease	125
Climate Change and Cancer	128
Climate Change and Kidney Disease	132
Climate Change and Liver Disease	135
Climate Change and Gut Health	140
Climate Change and Obesity	143
Climate Change, Dementia and other Neurological Diseases	146
Climate Change, Bone Health and Fractures	151
Climate Change and Dental Health	154
Climate Change, Fertility and Pregnancy	155
Climate Change and Eye Diseases and Blindness	158
Climate Change and Insomnia and its Consequences	164
Climate Change and Vulnerable Populations	167
Climate Change and Mental Health	171
Climate Change and Exercise	174

CONTENTS

Climate Change and the Pharmaceutical Industry   177

Climate Change and infectious diseases   **179**

   What is an infectious disease?   180

   Vector-borne diseases   182

   Why does Climate Change increase the risk from Infectious diseases?   183

   What Infectious diseases are we potentially facing?   185

The End   207
The Hummingbird   210
References and Notes   211
Climate Charities and Organisations   251
Further Reading and Viewing   254
Acknowledgements   257

INTRODUCTION

This book is written because not enough people are listening.

Our world is warming, and humans are to blame, but to convince you of this is not the primary purpose of the book. Dire climate warnings have become just noise on the TV and on news outlets or social media. Some piece of ice melting, some forest being chopped down, some wildfire spreading, some people on bikes wearing masks because of smog. But these are sounds and images of people that aren't you and those you have never met. You haven't witnessed the sea level rising – polar bears will adapt, more people should get air conditioning, you love the hot weather. It's not that you don't believe our Earth is warming, it's just that you don't see how it affects you.

Unfortunately for us all, it will affect you and me and indeed has already started to do so (whether you have or haven't noticed it). This book is about the real effects it will have on you via the thing you most care about – your own health. If anything makes any of us sit up and listen it is when our own health is at risk.

When you have finished reading this book you will realise – beyond doubt – that your health is threatened by climate change. This may or may not make you want to do something about our warming planet before it is too late, but at the very least, you will know what not acting will mean for you. In 2009

The Lancet and *The Institute for Global Health Commission* described climate change as the biggest global health threat of the 21st Century. That warning was over 15 years ago and the only thing that has changed since is that the threats to our health are accelerating.

This book is not about the solutions to the climate change catastrophe as there are plenty of books, podcasts, films and websites that will tell you what you can do. Neither will it berate you about why you are not doing enough to save the planet. Like smoking cigarettes, whether you decide to act is entirely up to you, but it is important you understand the personal consequences if you don't act.

I will say one thing though – if you think that our current attempts to reduce the rise in global temperatures are enough, then you are wrong. All reputable predictions indicate that we will not, if we continue emitting as we are, limit the Earth's temperature rise to less than 2°C. An above 2°C rise really does become an existential threat to humanity.

The book is not in-depth description of the science of climate change. To understand some of the impacts upon human health an understanding of the science will sometimes be necessary. There are many other excellent books about the science of climate change and I have mentioned some in the suggestions for further reading section at the end of the book.

Early humans were very much at the mercy of their environment. Every aspect of their lives was planned, altered and affected by the seasons, the weather and by natural events. As we humans have strived to bring the environment under our control this has resulted in changes to the diseases we have faced. Static, agricultural communities mean greater protection, food and water supply, but they also increase the risk of infectious diseases due to a greater mass of people in closer proximity. We have accepted this risk and increasingly have used the tools at our disposal to reduce them e.g. sanitation, vaccination, antibiotics. The threat of climate change is now at risk of undermining these efforts and putting our health advances back years.

We (in the more economically developed world at least) are now so powerful that we can control a large part of our immediate environment – we can move away from the coasts, we can fly from hurricanes, we can clear whole forests for agriculture. But this control has a price to pay – the energy we take from carbon sources is now beginning to harm us. Our climate contains so much extra energy and power that we are again heading towards a situation where we can't control its effects and are at its mercy. The more we go on trying to exert control, the more CO_2 we produce, and the worse things get. We have created a destructive spiral.

It struck me, when writing this book, that for all our mastery of the planet, we are moving back in time. As we will see, 'old diseases' that we thought were in the far past – leprosy, plague, cholera – are making a comeback. Those in the prosperous West at least, think of those only from our history books – the black death, leper colonies, cholera outbreaks – though in fact they hadn't disappeared they just affect places that the media doesn't usually reach. Climate change will supercharge these diseases of the past, allowing them to make a resurgence back into all our daily lives. As the world gets hotter, greenhouse gases rapidly increase, temperatures and CO_2 levels rise, and for all our progress, we paradoxically march back in time. We will pass back into the time before antibiotics, we will pass through the great plagues, through the times before any civilisations appeared, into times when methane dominated the atmosphere of a world with no ice and high, stormy oceans and no humans.

I had considered calling this book 'The Four Horsemen', referencing the four horsemen of the apocalypse in the *Book of Revelation*. They appear at the end of the world, representing pestilence (plague), war, famine and death. Runaway climate change does indeed cause an increase in infectious diseases, crop and animal destruction, conflict over land and resources and ultimately more deaths. But I decided to leave it up to the reader to make their own mind up if this would fit with the information in this book.

Anthropogenically (caused by humans) driven climate change has been called a 'Super Wicked Problem'. This term is used for a number of reasons: that time is running out, it has so far been impossible to stop, those who are responsible for it (you and me) aren't acting to prevent climate breakdown and governmental action is either weak or irrational. If you are looking for proof of this irrationality the recent terrible Los Angeles wildfires provide this with a leader who calls climate change 'a scam'. These fires also show that climate change will affect us all and neither money nor status can protect you.

In the recent United Nations Climate Change conference (COP) it was reported by the BBC that the Chief Executive used the conference to arrange deals with fossil fuel companies. At the same time a letter was published from 20 climate experts and leaders during the conference, stating that "COP is no longer fit for purpose". Our opportunities to avoid the disaster we are heading towards are not slipping away – they are being actively pushed away.

The Earth though, unlike its very recent human inhabitants, is used to Super Wicked Problems such as meteors, earthquakes and massive volcanic explosions. As the Gaia theory of James Lovelock details, the planet has dealt with them in the past and is already, as you will see in this book, starting to deal with the Super Wicked Problem that is humanity. Just like our bodies, the planet keeps the score.

It has often been said that the Earth will survive climate change but it's human inhabitants that won't. If you want to know why this is, this book will tell you.

Chapter 1

Is the Earth getting warmer (and why)?

Yes it is. Many confuse this indisputable fact with not *wanting* it to be true.

The facts that show our climate is undoubtedly changing are documented below. If, after reading these facts, you still don't believe that the Earth is warming or dispute that it is warming from human activity then do not read this book any further.

Please note the use of the word 'fact' in the first paragraph. A fact is defined by the *Oxford English Dictionary* as 'the truth about events as opposed to interpretation'. A fact is the truth because it can be proved. Facts are free of subjective bias – politics, religion, self-interest or whichever from the long list. Facts exist within their own universe – but they are not necessarily fixed, they can be changed, but **only** with new **evidence**. The modern world often appears to be anti-facts with many denying the truth – especially when it is personally inconvenient.

This book is not about those people or concepts. It is about what will happen to humans as the planet warms. We will only discuss facts – not opinions or interpretations – but facts. How do we know that things that will happen in the future are facts? Because we already know what is happening and the future can only be a more extreme version of this.

The boring numbers bit

Any book or article about climate change must have lots of numbers. Most people though are not comfortable thinking in numbers. As a list of warming indicators is read, we begin mentally to switch off – unable to see how a 40% reduction in coral reefs or a 50% loss of wildlife or a 2.5°C rise in temperature might have any meaning for us. This perhaps is one of the reasons we are not reacting to the climate crisis as quickly as we should (i.e. with a sense of urgency). Reading a list of numbers is not the same as watching a glacier melt or running from a wildfire or not having enough to eat. We can only really perceive something as real when we experience it ourselves. I shall resist making the point about the frog in the heating saucepan.

However much our difficulties are with numbers they remain vitally important to the description of climate change. If we are to talk in terms of facts, then we can't get away from using numbers as they are efficient and precise descriptions of those facts. If we are to know (as a fact) that the Earth is warming, then we need to measure it now and compare with past measurements. Without these comparisons we cannot *know* the Earth is warming. The rest of this chapter will describe these measurements and why they make us so sure that we are heating up.

Please don't skip this section as it is the very bedrock of the rest of the book. Do remember that all these numbers are based on peer reviewed, externally validated data. They are not opinions on *X* (formerly *Twitter*) or *Facebook*. I promise I will keep it as short as I can, but hopefully still convey all the information to convince you that the Earth is indeed warming and at an unprecedented rate.

If you actually like numbers or are interested to see how they relate to the climate, I recommend a paper by Santer *et al* in *Nature Climate Change*. The basis of any physical measurement could be a chance one. To prove it isn't, you must keep measuring until the same result occurs enough times for it to be reliably called a fact. Physicists call this threshold the 'five sigma level' which essentially means any other result is 3.5 million times less likely. The Santer paper uses this principle and feeds in climate data to show that human-driven climate change has reached the five-sigma level meaning that there is a one in 3.5 million chance that our climate is changing because of some other reason than human activities.

You can precis this finding as 'it is us'.

Thermometer readings

The oldest continuous temperature record is the Central England Temperature Data Series, which began in 1659. The Hadley Centre has measurements beginning in 1850, but there is too little data before 1880 for scientists to estimate average temperatures for the entire planet. Using 1880 as a starting point, the Earth's average temperature has increased by 1° centigrade (1.8° Farenheit) since 1900.

Unlike in 1850, this data now comes from many thousands of measurements over the Earth's surface. These show that 2010-2019 was the warmest decade since the records began. 2023 was the hottest year in human history. It saw the highest temperature ever recorded in China (52.2°C, 126°F) and the highest **midnight** temperature ever recorded (48.9°C, 120°F) in Death Valley, California.

I could list many many more numbers from all over the world but I won't because these numbers will be just that – figures that, in repetition, become abstract. Temperature records are being broken and by large margins in summers and winters and day and night. Deaths directly from the heat are continuously reported – especially in outdoor workers. Sporting and cultural events are increasingly interrupted by performers and audiences and equipment overheating.

It is impossible to say that our Earth isn't warming.

Indirect temperature measures

As you can see from the above, we have only been able to record accurately the environmental temperature for under 175 years. So how do we know that the temperature before 1850 (the lifetime of our great-grandparents) was as high or higher than today? Perhaps we are just in a natural temperature cycle that will correct itself? The answer comes in two main forms – tree rings and ice cores.

Both provide a 'snapshot' of previous climate events. Trees lay down a layer of wood each year and in this layer, substances can be detected that the tree has breathed in during that year. Similarly, as ice forms on top of ice, atmospheric particles are literally frozen in the layer they landed. Thus, both these mechanisms allow us to date and measure changes in the climate.

When we use these measures as comparators, we see that 1989-2019 was the warmest 30-year period in 800 years. Global surface temperature has increased faster since 1970 than in any other 50 year period over at least the last 2000 years. The Earth's temperature has risen by an average of 0.14° Fahrenheit (0.08° Celsius) per decade since 1880 so about 2°F in total. The rate of warming since 1981 is more than twice as fast: 0.32°F (0.18°C) per decade. 2022 was the sixth-warmest year on record at 1.55°F (0.86 °Celsius) warmer than the 20th century average of 57.0°F (13.9°C) and 1.90°F (1.06°C) warmer than the pre-industrial period (1880-1900). The 10 warmest years in the historical record have all occurred since 2010.

Evidence from other observations

Just taking the outside temperature is obviously of crucial value but it is not the only indicator of a warming planet. A wide range of other observations provide a more comprehensive picture of heating throughout our climate system.

Multiple temperature readings from multiple locations around the world and at all depths have measured warming of the oceans. Sea surface temperature increased during the 20th century and continues to rise. From 1901 through 2020, ocean temperatures rose at an average rate of 0.14°F per decade. Sea surface temperatures have been consistently higher during the past three decades than at any other time since reliable observations began in 1880. Based on the historical record, increases in sea surface temperature have largely occurred over two key periods: between 1910 and 1940, and from about 1970 to the present. The difference being that the 1910-40 temperature rise stopped whilst the 1970 rise continues and appears to be accelerating.

Satellite observations indicate that the Greenland and Antarctic ice sheets are rapidly melting. This loss has increased by 500% since the 1990s and now accounts for a quarter of world sea level rise. Since 1992, when satellite records of ice-sheet melt began, the polar ice sheets have lost ice every single year with the highest rates of melt occuring in the past decade. The Ice Sheet Mass Balance Intercomparison Exercise found that between 1992 and 2020, the polar ice sheets lost 7560 billion tonnes of ice – equivalent to an ice cube measuring 20km each side.

Melting ice and warming water increase sea levels all over the world. Low lying areas are at risk of floods and some islands are even at risk of disappearing altogether. Coastal areas have lives and livelihoods put at risk as floods destroy previously fertile soil. Polar ice disappearing might appear to be so far away from you that it is not a real threat. But it is.

Lived experience

If you remain suspicious of numbers or see them as abstract rather than representations of real events that have an effect upon you day-to-day, you will be no less concerned. During 2021-22, extreme weather conditions caused devastation across every continent. Floods in Brazil, China, Australia, Malaysia, Pakistan, South Africa and South Sudan caused thousands of deaths.

If you haven't been near a wildfire, then you will surely have witnessed some very frightening pictures on the daily news broadcasts. In Australia, in January 2020 wildfires burnt for weeks, destroying 12 million hectares of bush and farmland. 33 humans died and it is estimated a *billion* wild animals lost their lives. Few Australians who were safe in the cities could fail to notice weeks of smoke hanging in the sky and the magnificently frightening sunsets.

Choose a country or city by the sea and watch the tides increase in strength and depth. Flooding events that were once a century become once in a decade or less.

More mundanely you might have found your annual ski holiday more expensive as the snow retreats up the mountains. Or your summer holiday ruined by temperatures so high you couldn't sunbathe.

Part of a natural cycle?

Those sceptical of human induced climate change usually cite 'natural climate/weather/temperature variations' as the reason for the worrying statistics outlined above. If pushed to explain these natural cycles further, most of the sceptics would say variations in the Sun's output is the cause.

The Sun of course provides the energy that drives the Earth's entire climate system, so it seems superficially reasonable to blame its variability for the current situation. If we are in the time period where solar radiation is much higher, the Earth's temperature will obviously be higher. All we have to do is to wait for the Sun to calm down again. Unfortunately, there is no evidence for this. As the highly reputable *Climate Change Evidence and Causes* (update 2020) from the US National Academy of Sciences and the UK Royal Society states:

> For periods before the onset of satellite measurements, knowledge about solar changes is less certain because

the changes are inferred from indirect sources – including the number of sunspots (these indicate area of increased solar activity) and the abundance of certain forms (isotopes) of carbon and beryllium atoms, whose production rates in Earth's atmosphere are influenced by variations in the Sun. There is evidence that the 11-year old solar cycle, during which the Sun's energy output varies by roughly 0.1%, can influence ozone concentrations, temperatures and winds in the stratosphere. These stratospheric changes may have a small effect on surface climate over the 11-year cycle. However, the available evidence does not indicate pronounced long-term changes in the Sun's output over the past century, during which time the Earth's surface temperature has risen.

There is even more convincing evidence that excludes solar variability as the culprit for our global heat increases. If the increased energy from the Sun was to blame, then **all** levels of the atmosphere (from the Earth surface to just before Space) would be warmer. In fact measurements show that only the lower atmosphere (the troposphere) has warmed. Temperature readings from weather balloons and satellites show that the upper atmosphere (the stratosphere) has not warmed. This temperature difference is entirely as would be expected from increases in Carbon Dioxide (CO_2) in the lower atmosphere rather than solar activity.

Some of the other reasons given to avoid facing the truth of human induced climate change can also be discounted. Volcanic activity (which does increase CO_2 in the atmosphere) has not been substantial enough in the last 100 years to alter the climate. The two major natural climate variations El Niño and La Niña are also not consistent with the global temperature rises that are being recorded. Milankovitch cycles is the name given to the variations in the orbit of the Earth around the Sun that also affect the climate. This effect however, has been shown not to alter the climate significantly enough for the climate changes measured in the last 150 years.

The boring science bit

Understanding the science behind global warming is essential to appreciate why it is happening and why it is so important. Many people, as with numbers, feel uncomfortable with science. This book, however, is not about the science of climate change. I will cover the basics of this science as it is logical and understandable by those without a science background.

Why is the Earth getting warmer?

Carbon Dioxide was mentioned for the first time in the previous section. CO_2 is an invisible gas that surrounds us in tiny quantities (0.04% of the air) but has a profound effect. It is essential for life on Earth – not least in use by plants who need it for photosynthesis.

Around the mid-1880s some scientific experiments began to unravel the complex structure of our atmosphere and why it allows life to exist on our planet. The purpose of a glass greenhouse is to allow heat in through the glass but prevent the heat from getting out, hence the greenhouse warms and allows for more productive plants. The 'greenhouse effect' of a planet means our planet is warmer than it would be without an atmosphere. It therefore makes the Earth habitable for life. Heat from the Sun passes through the atmosphere, is then radiated from the surface of the Earth but on its way out is absorbed by gases such as carbon dioxide and water vapour. This means the heat cannot escape back into space. Further the CO_2 and water vapour radiate the absorbed heat **back** to the Earth. This results in an increase in the average temperature of the planet's surface, thus making our planet habitable (rather than a ball of ice).

The CO_2 and other atmospheric gases (greenhouse gases or GHGs) are acting like the glass of the greenhouse – letting heat in but preventing it getting out. Whilst this allows us to survive on our planet, too much CO_2 in the atmosphere means too much heat is trapped which then warms the surface of the planet.

This vitally important physical property of CO_2 was discovered in experiments by Foote and Tyndall (working separately) in the 1850s. Svante Arrhenius calculated in 1896 the temperature rise that would be caused by a doubling of carbon dioxide levels in the atmosphere. Guy Callendar showed in 1938 that human activity was responsible for increasing atmospheric carbon dioxide levels, and hence of potential global warming and climate change. We have known the dangers of too much CO_2 for 150 years.

Carbon Dioxide Numbers

The section above explains the importance of atmospheric CO_2 and its essential role in making our planet liveable, but whose same vital properties are now turning against us. The explanations above were based upon laboratory experimental and theoretical calculations. How do these fit with real world measurements and observations?

Direct measurements of CO_2 in the air and in air trapped for decades in ice, show that atmospheric CO_2 increased by more than 40% from 1800-2019. Most of this CO_2 increase has taken place since 1970 (which is consistent with the acceleration of global energy consumption at that point). Measured decreases in the fraction of other forms of carbon isotopes and a small decrease in atmospheric oxygen concentration show that the rise in CO_2 is overwhelmingly from the combustion of fossil fuels (that is coal and oil and gas removed from the ground).

By drilling into deep ice at the North and South poles, we can see what was happening in trapped air from past millennia. When this trapped air is analysed for CO_2 concentrations, it shows that currently the level of CO_2 is the highest it has been for 800,000 years. The earliest evidence of *Homo Sapiens* is from approximately 300,000 years ago, thus we are creating a world that we have never previously existed in.

As previously mentioned, CO_2 concentrations in the atmosphere are very small and therefore are measured in parts per million (ppm). But even at these tiny levels, CO_2 has a profound effect upon global temperatures. This is why the recent rises are so concerning. For many hundreds of thousands of years, atmospheric CO_2 stayed within the range of 170-300 ppm in the air. Over only the last 200 years this has risen to 400 parts per million. The last time CO_2 levels approached 400ppm was about 3-5 million years ago, and it is estimated that this raised global temperatures by 2-3.5°C. We will see later in this book what temperature rises at this level will do to human health.

As if a further warning was needed, 50 million years ago atmospheric CO_2 reached 1000 ppm. Little ice existed and not at the North and South poles (ancient bees have been found deep in the polar ice). Sea level was 60 metres (197 feet) higher than now. As you read this, think how high you now live above sea level and whether you would currently be underwater were the water at this height.

How do we know that the rise in CO_2 is caused by humans?

I will start by saying that the rise in atmospheric CO_2 that causes global surface warming is caused overwhelmingly by human activity especially that of the burning of fossil fuels. The evidence for this would fill a book by itself, and you could read such a book or simply search on the internet. Unless you solely search out the more extreme opinions (usually related to vested interests) you will soon be convinced. To save you time, I will summarise the evidence below – both the evidence for man-made changes and the evidence against other (natural) causes.

- Different forms of carbon have different 'fingerprints' i.e. it is possible to tell where they have come from. The form of carbon released by human activities is by far the commonest form of carbon in current atmospheric CO_2.
- Other gases that can cause a greenhouse effect (methane and nitrous oxide are the main accomplice culprits) show a similar human fingerprint.
- Natural variations in CO_2 output (volcanoes, solar activity, El Nino and La Nina variations) do not have their fingerprints in the CO_2. Sun activity varies over an 11-year

- cycle, but global warming has relentlessly increased despite a number of these cycles.
- Mathematical models indicate that natural variations – even at their most extreme – could not be responsible for the CO_2 levels that are currently measured. Natural simulations do not result in surface warming but in slight cooling.
- Conversely, when human influenced CO_2 emissions are put into the model the current level of CO_2, and hence surface temperature, fits the model almost perfectly.
- As was described previously, purely solar induced warming would only result in a heating of the upper atmosphere. CO_2 warming occurs in the upper and lower atmospheres of the Earth.

The speed of global warming is more than ten times faster than that at the end of the last ice age. Once again, this speed of warming is unprecedented in the Earth's history.

Feedback cycles

One further thing to mention is the importance of feedback cycles. Feedback cycles are used by the planet to regulate itself and take many forms. Suffice to say they are utterly crucial to allow life to exist on this planet. Disruption of any of these delicate cycles can have disastrous consequences. Unfortunately these disruptions, all because of fossil fuel burning emissions, are getting more and more common and will, unless stopped, eventually make our planet uninhabitable for humans and indeed for most animal and plant species.

The carbon cycle is the taking in and release of CO_2 by plants. Deforestation releases carbon from the dying trees and removes their influence from the absorption of CO_2. When this is added to the carbon from the burning of fossil fuels, the additional CO_2 disrupts the balance of the carbon cycle. This extra atmospheric carbon is too much for plants to take in, hence it remains in the atmosphere so increasing CO_2 with the consequences we have already seen.

As the surface of the planet heats, the ice sheets melt more and more quickly. Some of this ice has been frozen for millions of years. The ice traps CO_2 from the air and from decomposing life forms. As this ice melts this CO_2 is released causing more warming and so more ice to melt.

Additionally, as the ice at the poles melts the planet becomes less white. Less white means less reflection of the Sun's light (and therefore heat) back into space. Consequently, even more heat remains in our atmosphere.

Although CO_2 is the main greenhouse gas driver it is not actually the most potent – this is reserved for methane (CH4). All over sub-arctic regions the ground is frozen (called permafrost). Decaying organic material in this frozen soil produces methane, which is trapped until, as the climate heats, the soil thaws and releases huge amounts of this methane into the atmosphere.

As the Earth's surface continues to warm, more evaporation occurs from lakes, rivers and oceans. This means more water vapour is in the atmosphere. Water vapour – as we saw previously – retains heat and the warmer air can contain even more water vapour. More heat is therefore trapped and the atmosphere is warmer so more surface evaporation occurs.

When you watch pictures of tornadoes in the USA or notice that the rain where you live now sometimes comes in much heavier bursts, you are directly witnessing the result of extra water vapour (and extra energy) in the atmosphere.

The oceans themselves are important absorbers and storers of CO_2, and to some extent can negate the rises in atmospheric CO_2 we are describing. However, as the atmosphere warms, it warms the oceans. As the oceans warm, they have a reduced ability to absorb CO_2 so there is more in the atmosphere. More CO_2 means more heat which warms the oceans…

Disruptions and eventual destruction of feedback cycles are one of the most frightening aspects of global heating. This process can very quickly get out of control because of the amplification when feedback feeds in the wrong, destructive direction. They are one of the reasons that scientists can be so confident that global temperatures will continue to rise. Even if, in the extremely unlikely event that we switched off emissions today, the CO_2 already in the atmosphere and the negative feedback cycles it is causing will inexorably lift our global temperature.

Chapter Summary

- Our Earth, by all measurements, is unquestionably getting warmer.
- Although we have only had reliable temperature measurements of this for 150 years, ice and tree data show us that the planet is at its warmest for millennia.
- This rise in temperature is because of an increase in greenhouse gases in our atmosphere – particularly CO_2.

- Although there are natural cycles in the temperature of the planet, these are nowhere near enough to cause the temperature rises we are seeing.
- There is indisputable evidence that the burning of fossil fuels is responsible for the greenhouse effect and therefore of the rising temperatures.

CHAPTER 2

What happens to Nature when the temperature rises (and why that's important to humans)

The bottom line of global warming is that there is increasing energy in the Earth's climate systems. As basic physics tells us, energy is never destroyed so the extra energy must be somewhere – either being released or waiting to be released. The extra energy in the world's weather systems is not good news for nature including that small, but very loud part of nature – humans.

This chapter is about the changes to the natural world when this extra energy is incorporated into its systems. Without an understanding of this we cannot understand the health risks that we are facing. Nature has a wide definition but in this context it is about natural systems, their delicate balance, how easily this balance is disrupted and the consequences of this disruption. Eons of physical and then evolutionary change have created these fine balances and their fragility comes from the fact that so many natural systems exist side by side. Few of these systems can collapse without having profound effects on other natural systems.

The most obvious manifestation of climate change is the weather. Less obvious, in our day to day lives, are the profound changes that are occurring in the world's oceans. The major inconvenience of a cancelled skiing holiday is a personal 'tragedy' but sits on top of much bigger issues. We humans might notice the enormous variety of creatures that we share the planet with – some cuddly, some revolting. Perhaps though, what fewer of us realise, is how important all these creatures are to the cycles of nature and how invisibly dependent we are on them.

When we begin to understand how climate change is affecting our environment, we begin to move from the abstract world of climate science to the effect in the real world. If you understood the science in the previous chapter you would likely be frightened and this chapter ratchets this up. We will start with that constant topic of conversation – The Weather.

The Weather

The weather and the climate are often used interchangeably both in common language and to some extent in people's understanding of warming. They are of course intimately linked, but it is important to use them correctly or confusion and therefore opportunities for denial occur.

The *Oxford English Dictionary* defines **weather** as 'The state of the atmosphere at a particular place and time as regards heat, cloudiness, dryness, sunshine, wind, rain etc'. It defines **climate** as 'The weather conditions prevailing in an area or over a long period'.

The weather can be thought of as what is happening to me at this moment – am I too hot, too wet, too buffeted or am I comfortable? The climate is what the weather will be tomorrow or next week or next month or, increasingly important, in the next decade(s).

All too often you will hear someone say 'global warming? But it's freezing today', and this is the classic confusion of weather with climate. The climate is heating (see facts in previous chapter) but that does not mean you are necessarily too hot today. Daily weather has too many other factors that affect it for it just to be determined by warming of the atmosphere. Weather is, on a day to day basis, a poor marker of climate change. Unfortunately though with longer droughts, more intense rainfall, higher wind speeds (i.e. manifestations of increased energy in the atmosphere) our daily weather will increasingly reflect the disastrous changes in our climate.

As we have seen, if things are to be compared they must be measurable and to ensure everyone is comparing the same measurements they must first be defined. Extreme weather events are defined as 'unlike 90% or 95% of similar weather events that happened before in the same region'. Remember that climate change does not solely create the weather as weather is a complicated recipe of latitude, ocean cycles, forest coverage, El Niño and La Niña systems, but it *intensifies* the weather. No particular weather on a particular day can be said to be the result of human induced climate change, but a series of days outside the norm for that season and region can be. And that is the pattern that is becoming more and more common throughout the world.

Heat and Heatwaves

A warming climate increases the frequency and severity of heatwaves, their intensity and their length. This is perhaps the most logical and understandable result of the greenhouse effect.

A heatwave is defined by the Intergovernmental Panel on Climate Change (IPCC) as 'a period of abnormally hot weather, often defined with reference to a relative temperature threshold, lasting from two days to months'. In practical terms it is actually more complicated than this at it depends upon where you live as definitions vary in different countries. The essential elements of a heatwave are the same though wherever you live:

- A persistently high relative temperature for that region at that time of year.
- Usually accompanied by high humidity.
- A temperature that does not drop significantly during the night – making the night uncomfortable for sleeping and starting the following day at a higher temperature.

As we will see in later chapters, humans are not well-adapted to heatwaves and the elements above can have serious effects on the human body.

The 2018 IPCC report indicates that these hot extremes have become more frequent and intense across most land regions since the 1950s. Cold weather extremes have become less frequent and less severe in the same period.

Droughts

Increased atmospheric heat leads to increased evaporation of water from the Earth's surface into the lower atmosphere. Increased evaporation dries up rivers and lakes and the hotter it is the more evaporation occurs. The longer the heat spell lasts without rainfall, the greater the evaporation until a drought occurs. Remember also that water vapour in the atmosphere retains heat very efficiently, so more evaporation means more atmospheric water vapour with an increased greenhouse effect and so even greater warming (another destructive feedback cycle).

Less immediately visible than reduced rainfall is loss of ice and snow in the mountains. This doesn't just mean more expensive alpine holidays but represents a huge threat to millions of people. In many areas of the world summer melting of ice and snow on mountains feeds the rivers. Major examples of this are the Himalayan glaciers which feed the great rivers of Asia (the Ganges, Indus, Brahmaputra, Mekong, Thanlwin, Yangtze and Yellow Rivers). If these rivers dried up, the land of many, many millions of people would be in severe drought.

These changes are already upon us with the Kathmandu-based International Centre for Integrated Mountain Development (ICIMOD) reporting from the Himalayas that:

- Ice loss was as much as 65% faster in 2010s compared with 2000s.
- 30% to 50% of glacial ice will be lost by 2100 at 1.5°C of warming.
- The region is expected to hit 'peak water' by mid-century, followed by shortages.

In other parts of the world, ice-fed rivers are in the same position and this will produce the same consequences. We are losing these vital supplies of water for drinking and agriculture. Once we lose this mountain ice it will not come back. A warming world will not cool again and allow these huge ice reservoirs to reform.

Wildfires

A particularly nasty consequence of drought is the drying out of organic material – trees, plants, grasses, dead material – making these materials ideal fuel for wildfires. Spontaneous fires are common in nature and are often therapeutic – clearing old and dead material to allow new plants to grow. These are called landscape fires and are short-lived and cover relatively small areas. Wildfires are much more intense, longer lasting and cover far wider areas. In nature these are often started by lightning strikes in summer thunderstorms, but increasingly human behaviour is causing these fires. From carelessly discarded cigarettes to portable barbeques to the barely believable purposeful arson, humans, as usual, continue their self-destructive battle with nature.

The sting in the tail of wildfires is that the burning releases huge amounts of CO_2 from the organic (i.e. carbon based) materials and so adds to the greenhouse problem. Some of these wildfires can burn for months – especially in peat areas where slow persistent burning might not be visible but releases enormous amounts of CO_2 into the atmosphere.

The Intergovernmental Panel on Climate Change Report in 2022 showed that from 1979 to 2022, wildfire seasons lengthened across 25.3% of the Earth's vegetated surface, resulting in an 18.7% increase in the mean length of the global fire season. Recent years have seen deadly wildfires in the Amazon rainforest, Australian bush and Siberian and Californian wildlands.

Record-breaking wildfires that occurred in Turkey, Greece, Russia and California in 2021 were linked to climate change. It was also noted that wildfires caused significant air pollution, loss of water, loss of wildlife and loss of human lives as well as releasing enormous amounts of greenhouse gases. As climate change increases the likelihood of catastrophic fires, this leads to a vicious circle of gas emissions, global warming and ever-escalating wildfires.

More locally, wildfires release huge amounts of toxic substances. From large particulate matter to microscopic particulates to a whole range of poisonous chemicals, all of which have a significant effect on health. We will discuss this in more detail in later chapters, but wildfires are one of the most frightening aspects of our warming world. They are a visible representation of heat but at the same time pump out invisible particles that directly threaten human and planetary health. When you see the next television report of a major wildfire you really are looking at the face of hell.

Floods

A major feature of climate change is much greater variability in the weather. Rainfall might be heavier and longer than normal or lighter and shorter, similarly with heat and wind and even paradoxically ice and snow. In the previous section, we saw that drought events are increasing all over the world. But the water that is not falling as rain can't stay in the atmosphere forever and eventually must go somewhere. As more evaporation occurs more moisture enters the atmosphere. The atmosphere is warmer anyway so can hold more moisture. This excess moisture directly absorbs more heat from the sun. Thus we have a highly energetic atmosphere full of water which then falls on land with an intensity, energy and volume that can be very destructive.

Flooding is inevitable in this situation as too much rain falls in too short a time for the rivers and land to absorb it. Flash floods occur when huge amounts of rain falls in a very short space of time causing rivers to overflow. These are the reasons you see the pictures on your TV of cars and houses and people floating away. Such flooding presents obvious and direct risks to life and limb. Heavy rain events have increased relentlessly since the 1950s.

Flooding events worldwide are becoming more likely not only because of more intense rain. The greater atmospheric warming and resulting water evaporation dries out the soil, and this makes it less able to absorb water rapidly. As humans remove more and more trees and plants it makes the topsoil less stable so that heavy rain can easily wash it away (search 'mudslides' on *YouTube* if you want to be horrified). Building on land, tarmacked roads and domestic gardens turned into car parks all further reduce the ability of the land to protect us from flooding.

Less dramatic, but equally damaging, is when flooding persists over weeks or months. Persistence of flooding destroys crops and kills livestock. Flooding leeches away topsoil and when the flood recedes the land is less able to sustain agriculture. Drought reduces the amount of water in rivers and allows any toxins (natural and man-made) to be concentrated in these rivers. Flooding allows these toxins to spill onto the land. Flooding also allows, as we will see in forthcoming chapters, many disease causing organisms to flourish and spread.

Which one is the worst Ugly Sister I wonder – drought or flood? The UK Government's Climate Change Committee describes who Cinderella might be:

> ...adaptation to climate change remains the Cinderella, still sitting in rags by the stove: under-resourced, under-funded and often ignored.

The Oceans

Global warming has profound effects on the world's oceans. In turn, the world's oceans have profound effects upon the climate. Feared feedback cycles – which we discussed in the previous chapter – are played out on a terrifying scale.

The sea is not just something to head for in the summer, it is one of the major reasons for life existing on Earth. The oceans generate 50 percent of the oxygen we need, absorb 25 percent of all carbon dioxide emissions and capture 90 percent of the excess heat generated by these emissions. The seas are not just 'the lungs of the planet' but also its largest 'carbon sink' – a vital buffer against the impacts of climate change.

Unfortunately we have now pumped so much CO_2 into the atmosphere that the ability of the oceans to correct the climate is rapidly diminishing. Their efforts to absorb this extra CO_2 and the excess heat are creating problems in the oceans themselves.

Sea levels

Global sea levels are rising due to the absorption of heat as heated water expands and so takes up more space. Melting of the ice from Greenland and the Antarctic and from mountain ice adds to this. Thus global average sea levels have swelled by over 8 inches (around 20.3cm) since 1880, and since 1980 have risen a staggering 3.2mm per year. The dominant factor in global average sea level rise since 1970 is human caused warming. Sea levels are undoubtedly rising and at an accelerated rate – and would now continue even if we didn't produce another molecule of greenhouse gas.

What does it matter though if the sea levels rise? Can't we just move up the beach a little further? There is a vast amount of information about the impact of rising sea levels on human activity and this book is not aimed at this detail – except where it directly impacts on human health. Sea level rises have profound effects upon coastal communities – many humans, for obvious reasons, live in coastal areas and so the effects are going to impact many millions of people. The main effects are below:

- The most obvious is that vulnerable coastal land will be lost.
- Flooding will become much more likely with all the personal and economic implications.
- Coastal erosion will accelerate.
- Salt water contaminates soil and prevents crops from growing. This soil remains unusable for many years.
- Those living near the oceans are more likely to be affected by extreme wind or storm events.

Ocean Acidification

Increased CO_2 has more effects than just warming of the planet. When CO_2 gas is dissolved in water it becomes an acid. Ocean measurements have therefore become more and more acidic (called, in scientific terms, a lower pH). The pH of the global ocean has decreased by 0.1 pH units (i.e. seawater is more acid-like) since the preindustrial period and by a much faster rate since 1980 (around 0.02 pH units per decade). These may seem like small numbers but when you think of the enormous size of the oceans for us to have created such a change is shocking.

But what does a more acidic ocean mean? Marine species with shells have their shells dissolved by higher acid. Thus oysters, crabs, lobsters and clams will be increasingly threatened. You can think of this in purely human terms – coastal livelihoods lost, loss of your favourite seafood restaurant or increasing prices of shellfish. The real impact will be much wider though and will affect all marine ecosystems and their food chains.

A further impact of ocean acidification is an alteration in the cycling of vital nutrients and elements through the seas. This will further impact the delicate balance of these seas and will have impacts on all living creatures. Some species will be lost altogether whilst others move to different parts of the ocean and an ocean desert will be created in that area. Some species may even thrive in the empty spaces left but these are not necessarily the species you want to thrive. The economic impacts of the above in countries whose economies rely on exporting fish will be profound. For those communities whose diet is mainly fish-based the consequences will be deadly.

Warmer seas

The sea covers 70% of the Earth's surface, and luckily for us (for now) these seas have a very high heat capacity. It absorbs

90% of the warming that has occurred in recent decades due to increasing greenhouse gases, and the top few metres of the ocean store as much heat as the Earth's entire atmosphere.

But, once again, there is a limit to how much heat the oceans can absorb and we are reaching that limit. This then adds to the feedback cycle of global warming – increased air temperature means increased sea temperatures means higher air temperatures.

As we have seen above, the effects of ocean warming include sea level rise due to thermal expansion. We will see in the next section how this also increases the number and power of hurricanes. But warmer seas do other, more hidden, but equally terrifying things. They increase the ice loss from Greenland and the Antarctic – creating yet another disastrous feedback cycle. Warmer seas cause the death of coral reefs, and so change the ecology of the seas profoundly. Non-native species invade new areas, threatening native species who haven't had time to evolve any defences. Even marine animal behaviour changes – there is evidence that sharks are more aggressive in warming waters!

Perhaps one of the most frightening potential changes (in a long list) is that of alterations to the ocean currents. Ocean currents act much like a conveyor belt, transporting warm water and precipitation from the equator toward the poles and cold water from the poles back to the tropics. Ocean currents are vitally important in regulating global climate by helping to counteract the uneven distribution of solar radiation reaching the Earth's surface. Without currents in the ocean, regional temperatures would be more extreme — super hot at the equator and frigid toward the poles — and much less of the Earth's land would be habitable.

Changes in temperature of the oceans caused both by global warming and paradoxically by the cooling effects of melted sea ice has the potential to change the major global currents and therefore weather permanently. Warm places become too warm for life; temperate places become too cold for life. Again, this book is not about the science of ocean currents (which is complicated and still developing) but if you are interested in the science there are many good sources, some of which I have listed in the Further Reading section. If you can't be bothered reading these sources then watch the movie *The Day After Tomorrow*.

Just as the surface temperatures can produce heatwaves, the oceans have their own marine heatwaves. These are described as 'anomalously warm events' and defined as a marine heatwave 'if it lasts for five or more days, with temperatures warmer than the 90th percentile based on a 30-year historical baseline period'. Needless to say, these marine heatwaves are getting more common (a doubling in frequency since the 1980s) and the temperatures they reach unprecedented in previous measurements.

The IPCC Sixth Assessment Report stated in 2022 that 'marine heatwaves are more frequent, more intense and longer since the 1980s, and since at least 2006 very likely attributable to anthropogenic climate change'. This confirms earlier findings in the *Special Report on the Ocean and Cryosphere in a Changing Climate* from 2019 which stated that it is 'virtually certain' that the global ocean has absorbed more than 90% of the excess heat in our climate systems, the rate of ocean warming has doubled, and marine heatwave events have doubled in frequency since 1982.

Oh, and these don't just make the seas warmer – coastal land areas are even warmer when the sea is warmer. Yet another negative feedback.

Hurricanes and Tornadoes

Cyclones are large masses of air that rotate around a centre of low atmospheric pressure. Hurricanes and tornadoes are probably the best known – and certainly the most violent – examples of cyclones. Both have great potential to threaten life and limb and livelihood. They are different though in their origins and effects. Hurricanes form over water, are generally hundreds of miles wide and can have a cycle where they can weaken and strengthen again. Tornadoes form over land, are much smaller than hurricanes and much shorter lived. Typhoons are the name for hurricanes that occur over East Asia. The adverse effects of any of these are obvious and well known.

Earth's warmer and moister atmosphere and warmer oceans make it more likely that the strongest hurricanes will be more intense, produce more rainfall, affect new areas and be longer lived. Additionally, sea-level rise increases the amount of seawater that is pushed onto shore during coastal storms, which along with more rainfall from the storms results in more destructive storm surges and greater flooding. Similarly with warmer air and warmer land, tornadoes will become more common, more violent and will appear in more places.

Evidence so far indicates an increase in the global number of these severe wind events over the last 40 years. Observational evidence has noted an increase in hurricane activity in the North Atlantic. It is likely (over 66% probability) that the frequency and intensity of the strongest tropical storms have increased over the last few decades. It is very likely (over 90% probability) that tropical cyclones are strengthening further north in the western North Pacific Ocean region. Over the United States, hurricanes progress more slowly over the land compared with the early 20th century and therefore their destructive effects last longer.

The evidence linking global warming to these cyclone events isn't yet as conclusive as for some of the other effects of warming. In the next decade, we will be able to confirm if there is a definite link. It is obviously important to know as human frailty is never better illustrated than by these titans of nature.

Plants and Agriculture

Plants and their cultivation are essential to human existence. Any threat to them is a threat to our own species. It is therefore comforting to see a possible upside of increased CO_2 in the atmosphere that involves plants.

During the day, plants take in CO_2, turn it into nutrients and give out oxygen. They need energy in the form of sunlight to do this and the process is called photosynthesis. Increased CO_2 in the atmosphere should allow plants to synthesise more nutrients and so grow more rapidly. Increased heat allows the chemical reactions within the plants to be more rapid and therefore the plant becomes more fertile.

Indeed experiments have shown that elevated CO_2 levels increase some plant growth by an average of between 21 and 28%. As a result, some crops such as wheat, rice and soybeans are expected to benefit from this increased CO_2 with an increase in yields from 12 to 14%.

Whilst this is a rare potential positive in our climate change future, unfortunately, things are not that simple. Plants need other factors for optimal growth and the principal one of these is nitrogen. Plants can't take nitrogen directly from the atmosphere; the bacteria that live on the roots of some plants extract and fix it for them. The bacteria then pass it to the plants (who provide carbon in return – so a true symbiotic relationship).

These bacteria, like all living creatures, have a temperature where they function optimally. If the temperature goes too high the bacteria become less efficient and trap less nitrogen, meaning less nitrogen is passed to the plant. Further, a warmer atmosphere initially allows greater chemical synthesis in the plant but if the temperature goes too high the enzymes that allow the photosynthesis reactions start to malfunction. Eventually, the plant passes less carbon to the bacteria who in turn fix even less nitrogen to pass to the plants. Yet another negative feedback system occurs purely because of global temperature rise.

Climate change also means growing seasons are longer and warmer – allowing greater plant activity. However, greater plant activity means more water taken from the soil. Thus soils become much drier which in turn stresses the plants and reduces their ability to grow. Longer, warmer seasons can also mean greater opportunities for pests to attack plants.

Plants are now adapting to raised temperatures by moving habitat to try to regain optimal growing conditions. Species that have evolved to certain climatic conditions are gradually moving north or to higher elevations where it is cooler. In the last several decades, many North American plants have been found to have moved approximately 36 feet to higher elevations and 10.5 miles to higher latitudes every 10 years. The Arctic tree line is also moving 131 to 164 feet northward towards the pole each year.

All the above is occurring with relatively small increases in global temperatures. Larger predicted rises (greater than 2 degrees) will have serious, direct effects on all plants. An increase in temperature speeds up the plant lifecycle so that the plant matures more quickly, it has less time for photosynthesis and consequently produces fewer grains and smaller yields. Drought and flood cycles are, unsurprisingly, also very damaging to plants as are cyclones. The increased threat of new insect pests is discussed in the next section.

As plant species move nearer to the poles or to higher altitudes to try to find their optimal growing temperatures, these new environments can also contain potential threats that the plants have not evolved to deal with. Conversely, these invading species may present a threat to the species already in that area. This movement of living things (including humans) to try to regain their optimal living conditions and the conflict with species already there is something that will become an increasing threat to all the inhabitants of our planet.

Finally, we should touch on grass, more specifically wheat. Wheat is the most planted and traded crop in the world. Over 750 million tons are produced worldwide per year with half going directly to feeding humans and the rest to animal feed. It is an essential crop for the daily existence of most humans and domestic animals worldwide, so any reduction could have catastrophic effects. It is literally and figuratively your daily bread. Here is the data:

- Wheat yield decreased by nearly 6% in the period 1980-2010.
- Rising temperatures and rainfall changes are estimated to reduce this by a further 6% per degree of global warming.
- This will be less severe in the (current) temperate regions but more severe in the tropics and subtropics (where there will be least resilience).

As we will discuss in a later chapter, starvation and malnutrition are direct threats to human health from global warming.

Animals (except humans)

It is a vital theme in climate change that the Earth and all those creatures who live on and in it have interrelationships that have evolved for eons. Disruptions of these relationships will have consequences that we cannot predict. The loss of some bird species in the Amazon you have never heard of is intrinsically sad, but who knows if some time in the future this loss might have an effect upon you? From the discovery of new medicines to the loss of predators, to the loss of oxygen, our fellow creatures interact with us in a way we can never fully know. A permanent loss of a tiny beetle is the loss of a member of *your* family.

This book is not primarily about the loss of non-human animals. If it was, it would, sadly, be many thousands of pages long. But it is important to later chapters to explain what is happening to these animals and what will happen if we continue to warm our planet. An excellent description of the threat to animal life of increased temperatures is contained in the book by Mark

Lynas, *Our Final Warning: Six Degrees of Climate Emergency*. I would urge you to read it.

Species are being lost at an accelerating rate and from a combination of factors. Heat alters and stresses their bodies, droughts and floods take a severe toll. Loss of plants on the Earth and plankton in the sea means the food chains become unbalanced and lead to starvation. A 2016 study that surveyed 976 species around the world found climate-related local extinctions had already occurred in nearly half of them. This included fish, insects, birds and mammals.

As with plants, migration of animals to climates more compatible with those they have evolved into is a vital but potentially dangerous adaption to climate change. It means moving to cooler areas either further away from the equator or at higher altitude. For fish, migration means moving to cooler but unfamiliar waters and leaving behind a degraded ecosystem.

Unfortunately, all these forced migrations (human, animal, plant or microorganism) have their own intrinsic problems both for the migrant and the host species:

- Changing habitat does not mean your food supply comes with you.
- Travelling long distances to find cooler climates is hazardous and has no guarantees that they will be found.
- Previously unknown predators might be waiting and you have no defence.
- When species become restricted to smaller areas a single event can wipe out that species.
- Seasons change and therefore your food supply might not be ready when you move.

- Cooler temperatures may not necessarily mean enough water or shelter.

Just moving to another place when your current one isn't safe is natural but is hazardous. The combination of more hostile environments and the dangers of leaving these environments explains why many animal and plant species are disappearing at an alarming and accelerating rate. As the human population expands relentlessly (for now) we compete for these same environments with the current inhabitants and put even more pressure on them and their systems.

Insects

One animal class that we have particularly decimated is that of the insects. More than 40% of insect species are declining and a third are endangered. The rate of extinction is eight times faster than that of mammals, birds and reptiles. The total combined mass of insects is falling by a precipitous 2.5% a year according to the best data available.

But before they go, some of these insects have (some literally) a sting in their tails for us. With a few exceptions, insects are ectothermic. This means they are reliant on external environmental factors, especially temperature, to regulate their internal temperature. Their internal temperature has a direct link with their rate of development, reproduction and activity. Climate warming has therefore allowed some of these insect numbers to expand. It has certainly allowed insects to live further and further north and this we will see, in chapters 4 and 5, has profound effects on human disease.

The agriculture we depend so heavily upon will also be affected by the arrival of new insects. Warmer winters and a longer growing season help pests, pathogens, and invasive species that harm crop vegetation. During longer growing seasons, more generations of pests can reproduce as warmer temperatures speed up insect life cycles and more pests and pathogens survive warmer winters. Native plants do not have time to evolve defences against these new predators and are particularly vulnerable to destruction.

One thing that should be obvious from the changes we are creating in the climate is that we don't know what we are destroying and what it may lead to. It would seem from the section above that the increased temperatures that allow insects to thrive presents a significant risk. In reality, insect numbers worldwide are in precipitous decline for various of reasons – even insects cannot withstand extreme variations in temperatures, extreme droughts, loss of habitat, alterations of seasons and previously unmet predators. In one study it has been calculated that 65% of insect species could go extinct over the next century.

We might superficially think this is a good thing – fewer insects means fewer pests to eat our crops and to pass onto us tropical diseases. But insects are as important to our survival as any other animal or plant. They pollinate plants – without pollination we wouldn't have crops or flowers. They maintain healthy soil and recycle nutrients back into these soils. They control pests – yes, some insects actually protect us from other more harmful insects. They are food for all manner of creatures from birds to fish to amphibians. In short, they make the Earth what it is just as the Sun and soil and water do. Without any of these, we would or could be living on a lump of dead rock.

Microorganisms

Microorganisms are hidden from our sight but their size is in inverse proportion to their importance. They consist of a myriad of species of bacteria and fungi and protozoa. They represent another whole kingdom without which we would not exist. They are the most abundant living species on Earth and date back to the very origins of life. To list all their functions is beyond the scope of this book but there is not a biological function they are not intimately involved in.

As well as being essential to our survival they also threaten it. A small number of bacteria, and to a lesser extent fungi and protozoa, have waged a war of disease on us since we evolved. This war has had victories and losses on both sides but if we lose our defences (e.g. antibiotic resistance) there will only be one eventual winner.

Microorganisms have a real effect on climate change but conversely climate change has an effect upon them. There is no better example of this than phytoplankton. The sea is full of these tiny, microscopic creatures. Despite being so tiny, they take up 50% of all global CO_2 so significantly reducing the amount in the atmosphere. Increased atmospheric CO_2, as with land plants, will allow increased phytoplankton activity and therefore they will take up more CO_2. However, this is only true if they have also the necessary nutrients. Nutrients in the oceans are reduced by increased temperature because the warmer water circulates less, so more nutrients sink to the seabed. Which of these two opposing factors – higher CO_2 versus lower nutrient levels – will dominate? Already remote sensing data has shown a global decline in phytoplankton between 1998 and 2012 – particularly in the North Pacific. This process has been described as creating an 'ocean desert'.

There are many more examples of sea and land microorganisms that can ameliorate or worsen climate change. Unfortunately, even those that can reduce CO_2 can only do so within a narrow band of temperatures and only if other conditions remain stable. Increasingly, the climate is moving out of these tight parameters.

The seas are a highly complex mixture of delicate balances of nature and we really cannot predict the changes that increased global temperatures will bring. Will the increased CO_2 mean more toxic microorganisms thrive? Will the food chain be fatally disrupted by these changes to the plankton? Will coastal fishing communities be decimated?

Perhaps you don't feel any of the above has a real effect on you. You might respond by eating (increasingly expensive) fish only on special occasions or you might go to your local swimming pool rather than swim in the sea. So you will 'adapt' to the situation. Unfortunately, microorganisms are too entwined in our existence and far too numerous for us to be unaffected by them.

We will see in Chapter 5 one example of this when we discuss specific examples of infectious diseases. Climate change will change the distribution and nature of microorganisms in a way we may not have defences for. Climate change is predicted to increase antibiotic resistance. Disruption of rainfall and seasons will alter the microorganisms in ways we cannot predict. Because of their relatively simple structures and rapid reproduction they are, of all the Earth's creatures, best prepared to adapt and survive climate change. This is good in some ways, but for those microorganisms that cause human disease it is another threat created by our warming planet.

Volcanoes and Earthquakes

There is perhaps no better illustration of the far-reaching destructive power of climate change than its effect upon earthquakes and volcanoes. Whilst perhaps hard to believe, these immense natural phenomena can be altered by the climate; they are proof of the intimate connectedness of all Earth's systems.

Earthquakes occur within the deep surface of the Earth called the crust. Think of this surface being made up of huge plates that sit side by side but can sometimes slip over each other and the result of this slipping we call an earthquake. Although these plates are constantly moving naturally, there is evidence that external factors can influence this. Large amounts of rain/flooding can push down on the plates and make them less stable. Similarly, cycles of flooding and drought can destabilise the plates as the weight of ground above them alters. There is already evidence that this is occurring with increasing numbers of earthquakes in some parts of the globe.

A stark illustration of the unwitting effects humans can have on the planet comes from, of all things, avocado farming. In the Michoacán state of Mexico, which is the source of half of global avocados, it has been calculated that nearly 4000 Olympic sized swimming pools worth of water are used every day on avocado farms. This extraction of groundwater from aquifers is so large it has been implicated in the increased number of earthquakes in that area.

Volcanoes that are next to water can be triggered to erupt by rises in sea level. The extra weight of water pushes the Earth's crust, exerting pressure on the molten rock inside the volcano (magma) and this pressure can only be relieved through the top of the volcano. There are also innumerable volcanoes that are covered with ice and these ice caps prevent eruptions. As the atmosphere warms and we lose ice, more volcanoes will be free to erupt.

These are only phenomena that we know about and can measure and extrapolate, but we can't know the full effects of the changes to the climate. We have never lived in a world with these temperatures so we cannot fully know what lies in wait. The rest of this book is purely anthropocentric – dealing only with climate change that will directly affect human health. This selfish view of the world we share reflects the fact that our activities have created an entirely new geological era which is now named by some as the Anthropocene.

Ecosystems

Finally in this chapter, it is important to mention the concept of ecosystems. An ecosystem (*ecos* or οἶκος is derived from the ancient Greek for 'house') is a geographic area where plants, animals and other organisms, as well as weather and landscape, work together. They may be very big (e.g. the whole planet) or very small (e.g. a rock pool after the tide has gone out or even in just a handful of soil). What all ecosystems have in common though is that every organism – either directly or indirectly – is dependent on all the others. A change in the temperature of an ecosystem will often affect which plants will grow there.

Animals that depend upon those plants for food and shelter will have to adapt to this, move to another ecosystem or perish.

The whole surface of the Earth is a series of connected ecosystems, but even the largest systems are an exquisitely delicate balance. It is maintaining this balance that makes them so delicate – a loss of one species, a change in the weather, a foreign predator introduced and the balance may be completely disrupted. Millions of years of evolution have created this myriad of ecosystems that, by the evidence of our very existence, have worked for us. Man-made climate change is affecting all of them, and it is happening too quickly for them to adapt.

Most of the health hazards we will discuss in subsequent chapters are related to disruptions in ecosystems. You and I are just as much in balance with our surroundings as a deer is in a forest and eventually, as we will see, just as vulnerable to change.

Chapter Summary

- Climate change means there is more energy in the Earth's weather systems.
- This results in more intense heatwaves, wildfires, floods, droughts and cyclones.
- The oceans are warming as well as becoming more acidic, and this has profound effects upon sea level, marine and human life.
- The organisms we share the planet with will also be affected by climate change, and this will, in turn, have consequences for humans.
- Our ecosystems are in a delicate balance and disruptions can have effects that we can't predict.

CHAPTER 3

What happens to humans as the temperature rises?

This chapter is about the human response to raised external temperatures. Warmer climates will present a number of very difficult challenges for our species. Amongst other things, new diseases will appear, old diseases will reappear, traumatic events and food insecurity will increase. But these topics are for the next chapter, this chapter is about *What happens to us when we get hot?*

How do we normally regulate our body temperature?

Human beings are very sensitive to alterations in their body temperature. Evolution has created systems in us that work only within strict temperature ranges. Once we go out of these temperature ranges, our systems quickly begin to break down and, without a reduction in temperature, we die.

As temperature is so critical to our functioning there are many body mechanisms for controlling it by either retaining or losing body heat – when it is too cold or too hot. How hot or cold we feel and how warm our skin seems are poor indicators of our true temperature. The true body temperature is called our *core temperature* and is of vital importance. This is the temperature of the internal organs – liver, kidneys, spleen, digestive system etc. It is this core temperature that dictates how well our bodily systems work and so has to be kept in a very narrow range, 37.0 +/- 0.5 degrees centigrade (around 97.7 Fahrenheit).

All over our body – from our skin to our deepest recesses – temperature sensors continually feedback to a part of the brain called the hypothalamus. The hypothalamus not only monitors the information coming in but is able to act to reduce (or increase) our core temperature. Any rise in core temperature sends the hypothalamus into a frenzy of activity to quickly reduce this temperature. But it only has limited methods of doing this and if the external environment is too extreme it quickly runs out of options – as we will see later.

How do we avoid getting too hot?

Our body temperature is determined by the balance between our internal heat production and by the temperature of the external environment.

There are a multitude of chemical reactions (our metabolism) constantly happening throughout our body that perform a myriad of jobs but all produce heat as a by-product. This heat is essentially the furnace – the central heating – that allows us to survive in a changing environment. Our metabolic rate – our internal heat system – varies depending upon our activities. Exercise considerably increases the metabolic rate therefore increases our core temperature. Illness can also raise our temperature, but this is a method for the body to try to defend itself against infection.

So one mechanism our internal thermostat – the hypothalamus – can employ to regulate our core temperature is to increase our metabolic rate when the air temperature drops. The external temperature is obviously outside the control of the hypothalamus but it has to compensate for it. This is relatively easy when the temperature drops – we shiver or jump up and down – but if the external temperature is rising the body only has a limited number of mechanisms for cooling down. One way would be to reduce the metabolic rate. This, however, will only work in a limited way as bodily chemical reactions obviously need to continue so other strategies are needed to dissipate the heat:

- Radiation. This, is as the name suggests, is the heat that we *radiate* into the air. Like a fire radiating heat. This is the most important mechanism for heat loss at around 65%.
- Convection. This is when *moving* gas or fluid takes the heat away. If you run water across a hot surface it will cool it down as the water absorbs and removes the heat. When we move, heat is conducted into the surrounding air, when we use a fan the more rapid air moving across our skin takes more of the heat with it. Similarly, if we pour water over our head this is absorbed by the water and removed. 10-15% of body heat loss is by convection.
- Conduction. This is when heat is lost when in *contact* with a colder surface. So lying on the cold ground allows heat to be dispelled. Heat is conducted away even faster when in contact with water – which is why we cool down

so quickly when immersed in cold water. Conduction normally (as when in air) accounts for 2% of heat loss.
- Evaporation. Sweat produced by the skin evaporates into the surrounding air. Unlike the three above this is an *active* process because the water in the sweat needs energy to escape from the skin surface. It therefore takes this energy (i.e. heat) with it as it evaporates. Evaporation accounts for around 20% of body heat loss – much less than radiation. However, evaporation is the most important of all these methods as it is the only one that can happen when the **external temperature is higher than body temperature**. All the other mechanisms only work when the external temperature is lower than the body temperature as heat invariably moves from warm to cool so if these were the only mechanisms we had we would soon fatally overheat. Evaporation is therefore the reason we can live in hotter climates, survive heatwaves and exercise. The latter is very important as exercise significantly raises metabolic rate and therefore core temperature. Cooling is not fast enough by any of the other methods, so we only have evaporation as a tool and during intense exercise this increases to 85% of heat loss. It is very important to note that for external temperatures above 34 degrees centigrade (93 degrees Fahrenheit) the **only** method of heat loss is by evaporation. The other methods cannot work and in fact heat is gained by the body at these air temperatures.

Unsurprisingly, most of heat loss occurs via the skin – 90% in humans (breathing is most of the remaining 10% – although it is 100% in dogs and cats). It is via the skin that the hypothalamus utilises the methods above to prevent our core temperature from increasing. When the external temperature is sensed as rising (the skin has a huge network of temperature sensors which act as an early warning system) the hypothalamus quickly swings into action using both its internal control levers and external levers via behavioural changes:

- The hypothalamus reduces the metabolic rate so body heat production drops.
- The small blood vessels in the skin are opened to allow more blood to flow to the skin and therefore for heat to be extracted from this blood by radiation and convection. This is why we look red when too hot – including during exercise.
- Sweating is encouraged by increased blood flow to the skin's sweat glands and an increase in activity by the glands themselves so producing more sweat, and therefore more heat is lost by evaporation.
- Behavioural changes that are initiated by the hypothalamus include reducing physical activity, removing clothing, avoiding contact with warm surfaces e.g. rocks, seeking shade, using a fan to increase convection or jumping into a cold body of water. Whichever is chosen obviously depends upon availability, situation, societal norms and personal preference which are controlled by higher brain functions than the hypothalamus.

The scope of some of these mechanisms do vary in different parts of the world. Those who live in warm or cold temperatures do undergo some limited body adaptions that allow them to be able to better withstand extremes of temperatures. These adaptions though will not be nearly enough for the temperatures that are expected in our near future.

Wet-bulb temperature

A less generally known but very important measurement of heat is that of the 'wet-bulb temperature'. The air around us contains moisture – suspended water molecules which we call humidity. How many water particles are in the air depends upon the temperature and air pressure and as these vary enormously over 24 hours and between different regions humidity also varies considerably.

The reason humidity is so important leads on from the use of evaporation as a cooling method. If the humidity in a location is reaching 100% there is then no more room for further water molecules so sweat cannot evaporate from the skin surface. If water is not able to evaporate from the skin it can't take its heat with it and so cannot cool us down. A very hot day, combined with 100% humidity and no wind, leaves the human body with only behavioural methods to cool. If these behavioural methods are not available or not utilised and the conditions persist then core temperature begins to rise and we will rapidly (within hours) die.

Wet-bulb temperature is so important as it is the combination of environmental temperature and environmental humidity. It is called 'wet-bulb' because it was measured initially by wrapping a wet cloth around a thermometer. The wet cloth simulates the humidity and therefore gives a much more accurate idea of how uncomfortable we will feel at a given temperature.

The vital importance of humidity is often overlooked. You will have noticed yourself on certain summer days (and especially nights) that you are very uncomfortable. You are damp and sticky from the sweat that does not evaporate, you are hot but can't cool down and are irritable and find it difficult to sleep. Wet-bulb temperature has therefore, as you might realise from the above, a much greater ability to harm us even at lower temperatures.

At external temperatures above 45°C (113°F) even with low humidity, human systems will begin to break down and death will occur if that temperature is maintained. With wet-bulb temperature this number is much lower at around 35°C (95°F). In older and vulnerable groups who have greater temperature regulation problems, system breakdown begins at lower levels of temperature and humidity.

When the heat gets too high

Heatwaves

Heatwaves do not have a fixed definition as they vary from one region to another – one country's hot day may not be classified as hot in a more equatorial country. The main and overlapping definitions are as follows:

- The IPCC use 'a period of abnormally hot weather, often defined with reference to a relative temperature threshold, lasting from two days to months.'
- The Heat Wave Duration Index is a heatwave that occurs when the daily maximum temperature of more than five

consecutive days exceeds the normal average maximum temperature by 5°C (9°F). The same definition is used by the World Meteorological Organization.
- A definition from the *Glossary of Meteorology* is 'A period of abnormally and uncomfortably hot and usually humid weather.'

In essence, a heatwave is a prolonged period of abnormally hot weather. Heatwaves that occur in the summer months are particularly dangerous to health and most of the heat related deaths worldwide occur during heatwaves.

The numbers

As we saw in Chapter 1, the Earth is getting warmer, this is harming us in all sorts of ways, and that human activity is to blame is a fact not a theory. I could list pages and pages of numbers that show this but this would be repetitive and counter-productive, so I will simply take a particularly worrying set of numbers from the Climate Signals organisation website:

- The frequency of humid heat so high it comes close to overwhelming the human body's ability to regulate its temperature has more than *doubled* in some coastal sub-tropical regions of the world since 1979.
- Since 1950, the number and duration of heatwaves worldwide has increased due to global warming. 'It is *very likely* that there has been an overall decrease in the number of cold days and nights, and an overall increase in the number of warm days and nights, at the global scale.'

- Summertime heat extreme – defined as three standard deviations warmer than the climatology of the 1951-1980 base period – now cover about 10 percent of global land area, compared to much less than 1 percent of the Earth during the base period.
- Many urban areas across the globe have witnessed a significant increase in the number of heatwaves, with the largest number of heatwaves occurring in the most recent decade studied, 2003-2012.

People die during heatwaves. They die in different ways, and this is not always seen or recorded as a heat-related death. However, the figures for excess deaths (calculated by comparing previous years in that population) following heatwaves are clear. The World Health Organisation (WHO) has calculated these as follows:

- From 1998-2017, more than 166,000 people died due to heatwaves.
- More than 70,000 people died during the 2003 heatwave in Europe.
- Between 2000 and 2016, the number of people worldwide exposed to heatwaves increased by around 125 million.

It is stating the obvious to say that these numbers are only likely to increase as our planet warms. More and more heatwaves of longer duration and higher peak day and night temperatures are now an inescapable part of our future lives. In the next section we will see why people become ill and sometimes die in heatwaves.

Hotter and hotter

What happens to the human body as we heat up (assuming we can't or don't prevent this heating)? We have seen earlier in this chapter that the hypothalamus in the brain detects rising temperatures and uses various mechanisms to reduce our core temperature. But what happens when these adaptations aren't enough – either because the temperature and humidity are too high for too long or we have medical problems that prevent our bodies controlling our temperature?

When we are no longer able to regulate our body heat it is called hyperthermia. Heatstroke is a term we often use but it specifically means hyperthermia caused by high external temperatures (rather than a faulty heat controlling system in our body). Untreated hyperthermia (whatever the cause) triggers damaging changes in the body that would eventually lead to death. Although these are a continuum of events they are classified into different descriptions of increasing severity.

For any individual there is no specific temperature where these changes will happen. It depends upon the humidity, what temperatures they are normally used to, their general health and their levels of activity. Extreme exercise/exertion can cause heat-related illness even within normal atmospheric temperatures.

Mild heat-related illness. This can also be called mild heatstroke (if caused by external heat) or heat exhaustion or heat stress.

Heavy sweating is obvious and the skin becomes very red as more and more blood is pumped to the skin to try to lose heat. Fingers and toes and ankles swell (heat oedema) because the blood vessels have dilated so much they begin to leak fluid under the skin. Muscle cramps occur as the fluid and salts in the blood get too low. A rash may appear on the skin ('heat rash') caused by blocked sweat glands that can't lose sweat into the atmosphere.

At this stage the body is at maximum effort to try to reduce our temperature but after some hours this effort begins to fail. As it fails and the core temperature rises the cells and tissues start to become damaged. As irreversible damage to our cells occurs we enter the next stage of hyperthermia.

Moderate heat-related illness. Dilation of the blood vessels is so great and fluid loss so high that not enough blood is getting to the organs. Poor blood supply to the brain causes dizziness, confusion and then fainting ('heat syncope'). Each of the major organs – liver, kidney, bowel, lungs begin to fail. This results in nausea and vomiting, breathlessness and diarrhoea as the large intestine fails. Confusion worsens as the brain begins to die and this can lead to bizarre behaviour including feeling you are very cold and trying to warm up. Sufferers can become aggressive and try to fight off anyone trying to help. This behaviour is another example of a negative feedback system being initiated. Perhaps a metaphor for the climate itself.

Severe heat-related illness. The body loses all ability to control its temperature. The dying organs cause widespread inflammation that causes even further damage to cells and tissues. Our muscles begin to dissolve and parts of them enter the blood stream and damage the kidneys. The person will be severely confused, may fit and then lose consciousness. Their blood pressure will be very low, their pulse rate very fast (if it can be felt) and their breathing irregular. If there is too little fluid and not enough blood going to the heart, a section of the heart will die – a heart attack. The blood itself can become much thicker and therefore gets 'stuck' in blood vessels – this can also cause a heart attack or a stroke or a lung clot. Usually this stage is not recoverable as organs have been too damaged and the body's delicate proteins have been directly destroyed by the heat.

50° centigrade (122°F) is the upper limit of temperature for the vast majority of Earth's organisms to survive. Milk only has to be pasteurised a little higher than this, to 60°C, to kill all bacteria. Humans when directly exposed to 50°C would not survive for longer than a few hours. Until recently a temperature of 50°C or above was considered a very rare anomaly. In 2018 Nawabshah in Pakistan saw the hottest April ever recorded on Earth as temperatures hit 50.2C. In India in 2016, the town of Phalodi registered 51°C. Basra in Iraq registered 53.9°C also in 2016 whilst Kuwait City and Doha have also measured over 50°C or more in the past ten years. There are a very small number of organisms that can survive prolonged temperatures above 50°C (called thermophiles) – a few bacteria and some much simpler single cell creatures. It looks like they will be the ones to inherit the Earth – or I should say, as we once again travel back in time, re-inherit the Earth.

The above is a summary of how excess heat directly and acutely affects humans. This is a situation that health care providers are increasingly having to deal with. There is evidence that heatwaves are already becoming more common. It has been shown that even very small rises in atmospheric temperature exponentially increase the number and intensity of heatwaves.

Heatwaves are an obvious threat to human life and we have seen the reasons why they are so deadly. What is less immediately obvious are the other threats to human health that will be unleashed with even a modest 2°C (35°F) global rise in temperature. A 2°C rise in your central heating would be barely noticeable but in planetary terms is a threat to our very existence. The next chapter will explain why this is so.

Chapter Summary

- Human temperature is maintained within very strict parameters to protect our bodies' cells and tissues.
- When our environment heats our brain works hard to maintain our core temperature.
- If the external temperatures continue to increase, we may not be able to prevent this core temperature from rising. High humidity significantly adds to this.
- This has very serious effects upon our organs and will, if we do not cool, be fatal.
- Heatwaves are becoming more common and more deadly in our warming planet.

Chapter 4

Threat by Threat

We are evolved and adapted to the Earth's climate as are all the living creatures we share the planet with. Climate change/global warming changes the Earth in predictable and unpredictable ways. It is therefore not the same planet anymore – it is not our home as we knew it.

This is not an aesthetic or philosophical point but has real practical implications. One of these implications is the changes in type and distribution of threats to our health. As we will see, there are very few areas of our well-being that aren't negatively affected by climate change. Our environment is so intimately linked to our health that changes in it can have profound effects on our well-being. Because evolution is so precise in its outcomes it leaves us stranded when our environment changes too quickly.

If you want to know what this means in terms of numbers, these are helpfully provided by the World Health Organisation who report that one in four deaths can be attributed to preventable environmental causes, with an estimated additional 250,000 people dying every year due to climate change. Rising temperatures, extreme weather events, air pollution, wildfires, and compromised water, land and food security result in lives lost and negatively impact infectious diseases, heat-related illnesses, noncommunicable diseases and adverse pregnancy outcomes. How much does this cost? The health consequences of climate change carry significant economic ramifications. The World Bank estimates that up to 132 million people will fall into poverty by 2030 due to the direct health impacts of climate change, and approximately 1.2 billion people will be displaced by 2050.

The preceding chapters have explained how our world is changing in so many ways in response to its warming. This chapter and the next ties all this together to outline the threats to human health. The crisis we now face is man-made – it is anthropogenic and these chapters are purely about the threat we have created for ourselves. As we have seen, the Earth and its occupants are intimately related, and if we change even one element we do not know what the outcomes will be. This chapter will describe the threats and risks we humans are facing.

Health threats on a warming planet

Direct effects to Human Health

Global warming, the Greenhouse Effect. These are causing, as we discussed previously, retention of more and more heat in the atmosphere. Heat is a form of energy and energy can never be lost – that energy must go somewhere. Eventually some of that extra energy will be heading towards you. To say that humans are puny in respect to nature's forces is a cliché but often all too true. This section is about the various direct forces that climate change will harm us with.

Earth

The most destructive forces (humans aside) on our planet are undoubtedly earthquakes and volcanoes. These colossal forces of nature are more likely to be unleashed with continued climate change. Already data has shown increased seismic and volcanic activity as our planet warms.

As we saw in a previous chapter, climate change is thought to increase the occurrence of earthquakes. The mechanism for this is heavy rains, higher sea levels and ice melting stressing unstable tectonic plates. There is some evidence that earthquake activity has increased over the last century and particularly since 1970. It is difficult to be certain about this as there is a variation in the number of earthquakes that occur each year worldwide.

The WHO reports that:

Between 1998-2017, Earthquakes caused nearly 750,000 deaths globally. More than 125 million people were affected by Earthquakes during this time period, meaning they were injured, made homeless, displaced or evacuated during the emergency phase of the disaster.

Thus, if global warming is causing – or is going to cause – an increase in number and severity of earthquakes, then the outcome for humans is dire. (I don't think we should wait for definitive evidence that man-made climate change is affecting the Earth's crust – when crossing the road would you cross when a speeding car is heading toward you on the assumption it will probably stop?)

Similarly with volcanoes. We have seen previously that these are likely to become more common as ice caps melt and sea levels rise. As with earthquakes, there is not yet firm evidence that the number of eruptions per year is increasing. There is some evidence though of an increase in smaller eruptions, especially in areas where ice is melting. (Again, do we really want to wait to have definite evidence that we are causing increased volcanic activity?)

Volcanoes kill in numerous ways, some local and some at far distances. These include an initial explosive event, lava flow, volcanic mudflows, poisonous gas release, lightning storms and burning ash. The latter can set anything it lands upon on fire. The ash can be ejected so far into the atmosphere that it lands hundreds of miles from the eruption. Ash from larger eruptions that stays in the atmosphere can reduce sunlight reaching the ground and therefore create a mini-winter. These volcanic winters have been documented in the past and some have had effects all over the globe. Ice core evidence shows a massive eruption around 71,000 years ago that expelled so much ash into the atmosphere that a mini ice-age was created. Some experts believe that this nearly made early humans extinct. I will not pursue the irony of our current position except to say that this time it would be our own fault.

Over 29 million people worldwide live within just 10km of active volcanoes, and around 800 million people live within 100km, a distance within which there is potential for the devastating volcanic hazards listed above. But as we have seen, even those at far distances from active volcanoes are at risk from the bigger eruptions.

Wind

Perhaps the most obvious, as in everyday, example of a direct natural force is the wind. We have all been battered by high winds at various times in our lives. Most of us can remember storms that rattled the windows and took the slates off the roof and brought down trees. These minor wind conditions will become more common and more destructive. Most of the destruction is economic – unlike their much bigger siblings.

Winds in their most severe form – hurricanes and typhoons and tornadoes – mean death and destruction are inevitable when land is hit. The US National Weather Service has noted over the last five years increased tornado numbers, earlier onset of the tornado season and an increased number of direct fatalities.

Marchigiani and co-workers provide an excellent summary of Wind Disasters (WD) in the aptly named *International Journal of Critical Illness and Injury Science*. The numbers are startling and even a small increase in them would have huge effects worldwide:

Between 1980 and 2008 worldwide:

- 1211 wind related events (i.e. Hurricanes/Typhoons or Tornadoes) were recorded.
- 402,911 people were killed.
- 496,560,637 people were adversely affected by WD.
- 500 million USD of estimated economic damage occurred.

Over the same period just in the USA:

- 34,438 wind related events were recorded.
- 4,114 people were killed.
- 70,034 were injured.
- 1.9 billion USD economic costs were sustained.

Tropical cyclones (Typhoons) are equally destructive, causing a 6% increase in deaths in the first 2 weeks following the events. They are estimated to be responsible for nearly 100,000 excess deaths per decade over the two weeks following them. This means over 20 excess deaths per 100,000 residents between 1980–2019.

The figures above clearly show a profound threat to human life and to the economies of the countries affected – wherever you live on the globe. Some of these countries are better equipped to deal with such ongoing catastrophes than others.

Wind events cause injuries in direct and indirect ways and the mechanisms can be divided into three overlapping phases:

Injuries prior to the impact. Increased death and serious injuries have been shown to occur prior to wind events. People trying to shore up their houses, fleeing in cars or moving vulnerable elderly people have all been shown to increase the rate of serious injury and fatalities.

Injury during the impact phase. This is caused by direct wind damage – literally knocking down. More commonly, injury is caused by being hit by debris propelled by the wind. Those in mobile homes are at particular risk (85.1 per 1,000 as compared to 3 per 1,000 for standard construction homes) but those in poorly constructed buildings are no safer. The storm surge of sea water that occurs just before and during WD causes significant numbers of deaths for those who live near the coasts.

Injury patterns tend to involve multiple systems. Commonly injured anatomic regions include the chest (45%), abdomen (27%), extremity (91%) and head (45%). Furthermore, trauma severity increases if the victim is thrown rather than struck by flying debris. Most of the serious injuries and deaths are the result of the victims or solid objects becoming airborne, or structural collapse. Death is most frequently attributed to head trauma, followed by crush injuries to the chest, abdomen, and pelvis. Most tornado fatalities die at the scene either in the open or in unstable structures.

Injury during the post-impact phase. Numerous risks persist even when the winds have passed. Falling objects and masonry, collapsing buildings and electrocution from ruptured power lines are amongst the most obvious. Less obvious perhaps is the increase in burns from makeshift candles and stoves with the latter also being responsible for carbon monoxide deaths when in a confined space.

As we will see from a later section, a very real risk to health is the overwhelming of healthcare faculties. The healthcare facilities themselves may have been damaged or their staff injured or evacuated. For example, in the wake of Hurricane Sandy in 2012, even several weeks after the hurricane three New York City hospitals were still closed. Initial estimates of healthcare-specific damages in New York alone were as high as $3.1 billion. This occurred in one of the richest and mostly richly provided healthcare cities in the world, so it takes little imagination to see the devastation that can be wrought in developing countries with poorly functioning healthcare.

In the longer term, these events have been shown to increase mental health problems and particularly post-traumatic stress disorder (PTSD). This is particularly so in those already at greater risk of harm from these wind events (as well as all climate related threats), namely the very young and the very old.

It is important to remember that economic impacts are not purely about money but can have real detrimental effects upon people's health. Local economies can simply collapse and on a wider scale government rebuilding money has to be taken from something else. As poverty is one of the major determinants of health and lifespan, the deleterious effects of catastrophic events such as wind disasters can continue through lives and generations.

Fire

Human beings have feared and harnessed fire for eons. We know we can't live without fire but we retain a healthy respect for it. Fires kill thousands worldwide every year. But the overarching risk of fire in our era of global warming is wildfires. The United Nations report *Spreading like Wildfire: The Rising Threat of Extraordinary Landscape Fires* is a comprehensive account of these threats.

As we discussed in Chapter 2, the risk of wildfires is increasing. The 2022 IPCC report quantified this. From 1979 to 2022, fire seasons lengthened across 25.3% of the Earth's vegetated surface, resulting in a nearly 20% increase in the mean length of the global fire season. Recent years have seen deadly wildfires in the Amazon rainforest, Australian bush and Siberian and Californian wildlands. Record-breaking wildfires that occurred in Turkey, Greece, Russia and California in 2021 were linked to climate change. As climate change increases the likelihood of catastrophic fires occurring, this leads to a vicious circle of ever-escalating wildfires and global warming.

Wildfires have direct and indirect effects upon human health. Direct threats are most obviously from the burns that cause death and injury. Wildfires can rapidly progress if there are also significant winds. Cruz and Alexander (2019) calculated that fire speed was roughly 10% of prevailing wind speeds – so that if the windspeed was 25km/hr (16 mph) the fire would be spreading at a rate of 2.5km/hr (1.6mph). Windspeeds of 80-100km/hr (60-50mph) have been recorded in these situations resulting in fire speeds up to 10km/hr (5mph) and this can be even quicker in grasslands. Firstly, just think of the enormous quantity of land destroyed during this time. Secondly, it might seem strange to be caught in one when you can obviously see a wildfire coming but at these speeds you may have little time to get clear of the danger. Hot ash and embers can also be carried by the winds, setting fire to houses some distance from the initial blaze.

Wildfire smoke can travel thousands of miles, potentially resulting in widespread adverse health impacts. The smoke from a 10,000-hectare wildfire can affect people living in an area 10 to 15 times larger, impacting people who have never even seen the flames. Most of the damage to health comes from breathing in the smoke with resulting respiratory problems.

For those already with lung diseases such as asthma or chronic obstructive airways disease (COPD) this smoke is particularly dangerous. A longitudinal study conducted in Brazil showed that a 10µg/m^3 increase in wildfire-related particulate matter was associated with a 1.7% increase in **all-cause** hospital admissions, a 5% increase in **respiratory** hospital admissions and a 1.1% increase in **cardiovascular** hospital admissions between 0 to 1 days after the exposure. Estimates from studies in 749 cities across 43 countries suggest that wildfire smoke and its associated air pollution directly cause over 33,000 deaths annually with an increased risk of all-cause mortality (i.e. it's not just the lungs it damages).

Wildfires produce huge numbers of tiny particulate matter (we will go into more detail about this in a later section) and these particles have been shown to reduce immunity – meaning that it is harder to fight off infectious diseases. The natural barriers that also protect us from infection in the gut and lungs are disrupted by the particles and this raises the risk of infections even further. Inflammatory markers are raised in sufferers, and this probably causes the increase in cardiovascular diseases. All these factors, yet again, are more likely to affect the young and the old as well as pregnant women and the foetus.

As with major events such as earthquakes and volcanoes, significant wildfires take their toll on healthcare systems. As well as potentially overwhelming healthcare facilities those very facilities might be damaged by the fire. Transportation to facilities might be impossible and this may also prevent healthcare workers reaching their workplace. We see once again climate change creating cycles of damage.

Once the fire has passed, the human toll does not diminish due to ongoing respiratory and cardiovascular issues. Mental health impacts persist with PTSD and can be particularly prevalent. And, of course, these fires come in seasons with no guarantee the region will have recovered by the time the next fire hits.

Wildfires can affect the water supply of an area, resulting in crop damage and drinking water contamination. This has various mechanisms. Increased pollution of reservoirs from wildfire debris. Fire damage shortens reservoir lifetime and increases maintenance costs. Loss of plants and trees to bind the soil make soil erosion more likely. This erosion means greater transportation of sediment and debris to downstream water-treatment plants, water-supply reservoirs, and aquatic ecosystems. Less water allows a greater concentration of toxic chemicals. More stagnant water allows microorganism overgrowth with production of toxins.

Economic damage from the wildfires can be severe and significant economic impacts tend to have significant effects on health and healthcare. The 2017 report from the National Institute for Standards and Technology found that 'the annualized economic burden from wildfires was estimated to be between $71.1 billion to $347.8 billion'. This is just in the USA, so the true worldwide number is much greater. Wildfire numbers have increased since 2017, so even in the US this number is an underestimate. It is often stated by critics of Net Zero (the amount of greenhouse gas that is produced and the amount that's removed from the atmosphere becoming equal – so that no more CO_2 builds up) that it is too expensive to get to. For those who fully understand the situation we are in, it is obvious that it will be far too expensive not to get there.

Finally, and this is something so important it is worth repeating from Chapter 2, wildfires push enormous amounts of CO_2 into the air. They therefore add to climate change and global warming. The warmer the climate becomes the more likely are wildfires. We create yet another disastrous feedback loop.

Water

The threats to human health from water issues are both obvious and more subtle. One of the major ways that climate change exerts its effects is by alterations in water supply and in changes to the quality of that water. No living thing on this planet can exist without water so it is, as many things are in climate change, an existential threat not only to us but to all our fellow organisms.

We will take the most obvious and direct threats from water first – floods and droughts. They are not opposites, as you might think, but close cousins that are equally affected by global warming.

Floods. As we discussed in Chapter 2, flooding occurs when the amount of water arriving in an area (e.g. heavy rain, tidal surge) is too great for it to be drained. This is related to the inability of the land to absorb the water either because it is being deposited too rapidly or because the land has been altered by drought, deforestation or the narrowing of rivers.

The IPCC is clear about the risk and dangers of flooding. Climate change has already made floods more likely and more severe. Consequently, flood risks and associated societal damage are projected to further increase with every degree of global warming. The frequency of heavy precipitation events will increase over most areas during the 21st century, with more rain-generated floods. Water-related disasters have dominated the list of disasters over the past 50 years and account for 70 per cent of all deaths related to natural disasters. Since 2000, flood-related disasters have risen by 134 per cent compared with the two previous decades. Most of the flood-related deaths and economic losses were recorded in Asia.

The most immediate health risk from floods is obvious – drowning. The WHO already describes drowning as an under-recognised threat to public health. Global estimates of the mortality burden of unintentional drowning are around 200-300,000 per year worldwide, though this is likely to be an underestimate. There were 3254 floods globally between 2000 and 2019, compared with 1389 between 1980 and 1999. Between 2000 and 2019, floods resulted in 104,614 deaths. The WHO estimates that drowning accounts for 75% of deaths in flood disasters.

As the ambient temperature increases, so does drowning in open water settings. Parks *et al* demonstrated that a 1.5°C anomalously warm year would be associated with a 13.7% increase in drowning deaths in men aged 15–24 years in the US. Among all injury types, drowning is the injury type most affected by increased temperatures. Warmer temperatures lead people to spend longer in the water and evidence from Australia highlights increased alcohol consumption on days with hotter temperatures. Remarkably, continued ice melt also increases the risk of drowning directly in areas where the ice is melting – in the areas of Canada with the greatest ice-melt, the number of drowning events is increasing.

High wind events create storm surges. A storm surge is an abnormal rise in seawater level during a storm and is measured as the height of the water above the normal predicted astronomical tide. The surge is caused primarily by a storm's winds pushing water onshore. In coastal areas these surges are responsible for far more deaths than from the winds themselves. Most drowning events occur, unsurprisingly, in low lying coastal areas in developing countries. Warning systems do reduce fatalities but unfortunately do not eliminate them. One cubic metre of seawater (at 20°C) weighs 1024 kg — over a ton. A storm surge carries tons of water at speeds typically from 15 to 25 kilometres an hour. This speed traps even those who have been warned and the power contained in such a weight and such a speed destroys everything in its path.

Flooding risks are not restricted to those areas near the sea. Rivers present significant risks if there is enough rain in the right conditions, such as very dry soil or previously rerouted rivers – especially through villages and towns. We can think of these events in two different ways with different risks for each. Flash flooding is flooding caused by heavy or excessive rainfall in a short period of time, generally less than six hours. Flash floods are usually characterised by limitations in the spread of the water such as narrowed or dried out rivers, urban streets or gaps between hills and mountains. The energy within these floods is significant so drowning events are not rare and damage to goods and property is very likely.

An even more destructive situation can occur when dams are overcome, and thereby reservoirs empty in a very short period of time. The force and volume of the water can cause huge numbers of casualties and destruction. In Libya in 2023, over 15,000 people were simply washed away by a catastrophic dam burst. Climate change has and will make dam bursts more likely for several reasons. Sudden intense rain events may prove too great for the dam to contain. Temperature increases causing geological changes (for example glacier melts, permafrost thawing) can shift the dam's bedrock and weaken its structure. Avalanches and landslides can damage the dam structure or again add too much water too quickly for the dam to hold.

Slow-onset flooding is obvious from its name. It is part of a set called 'Slow onset disasters' which are environmental degradation processes such as droughts and desertification, increased salinisation, rising sea levels or thawing of permafrost that occur over long time periods but are often cumulatively more deadly than 'Fast onset disasters'. This slow onset flooding is already becoming more obvious worldwide. It does not present significant drowning risks but is becoming more deadly. It destroys livestock and crops, and if seawater can make the soil unusable for future use thus starvation and malnutrition are vastly increased. The static water increases the risk of waterborne diseases (see next section) as well as increasing risks from venomous snakes, spiders and crocodiles. The floodwaters overcome the sewage systems and raw sewage enters the water supply. Similarly, toxic chemicals are washed into the water as is waste from factory farms. Once again, healthcare provision cannot stand separately from these events with the overwhelming of health resources, damage to healthcare facilities, damage to the transport of medical supplies and loss of staff.

A study involving over 700 communities from 35 countries was published in the *British Medical Journal* in 2023. This showed that mortality risks increased and persisted for 60 days after a flood event. In particular, heart and circulation and lung disease were responsible for most of this excess mortality.

Flooding presents a major risks to humans – from the early danger of drowning to the short term increase in heart and lung deaths, to the risk of infectious diseases to the long term spoiling of agricultural land.

Inevitably, there are also huge economic costs related to flooding events. Economics cannot be separated from the resources available for healthcare – even in the richest countries. The US Natural Resources Defence Council reports that between 2007 and 2017, the National Flood Insurance Program (NFIP) paid an average of $2.9 billion per year to cover flood-related losses, with individual years often costing far more. Following Hurricane Sandy in 2012, property owners filed approximately $8.8 billion in flood claims. Another $8.8 billion would be filed five years later, after Hurricane Harvey. Climate change will inconvenience all of us with higher insurance premiums – for example, the NFIP has been deeply in debt since Hurricane Katrina. Major floods are becoming more common and more costly in all ways.

The human and economic costs of flooding can perhaps be reasonably contained in developed countries but in the rest of the world probably can't. As we continue to warm, more and more deaths will occur and economic tolls will increase until, eventually, they are not affordable by governments. We will see in a later section just how important national economics is to the functioning of healthcare.

Droughts. Drought can be defined in different ways, but essentially it is a period of unusually persistent dry weather that occurs for long enough to cause serious problems such as crop damage and water supply issues. Droughts kill, slowly but very effectively. The World Meteorological Organization states that it has been estimated that droughts are the world's costliest natural disaster, accounting for 6-8 billion US dollars annually, and impact more people than any other form of natural disaster. Since 1900, over 11 million people have died as a result of droughts and 2 billion people have been affected. Since the 1970s, the land area affected by drought has doubled, undermining livelihoods, reversing development gains and entrenching poverty among millions of people who depend directly on the land. In the period from 1970 to 2012, drought caused almost 680,000 deaths, mainly due to the severe African droughts of 1975, 1983 and 1984.

The IPCC again report that climate change has made droughts more likely and more severe. Rising global temperatures increase the moisture the atmosphere can hold, resulting in more storms and heavy rains, but paradoxically, this also causes more intense dry spells as the raised temperatures means water evaporates from the land and global weather patterns change leaving some areas without rain. As with floods, droughts are projected to further increase with every degree of global warming. The proportion of land in extreme drought at any one time is also projected to increase. The number and duration of droughts increased by 29% over the period from 1970-2012 Most drought-related deaths currently occur in Africa, and this will continue but more and more deaths will take place in other continents.

There are various adverse health impacts of drought. Some are particularly more severe in some regions compared to others, but generally the higher the average temperature the worse the effects of drought:

- Many communities worldwide get their water supply directly from ground sources (streams, wells, aquifers). Even in large urban areas, many of the reservoirs are fed from rainwater and groundwater (and have the additional issue of temperature-driven evaporation increase). Some droughts only show up after years of reduced rainfall – but are just as deadly.
- Reduced stream and river flows increase the concentration of pollutants.
- Higher water temperatures in lakes and reservoirs lead to reduced oxygen levels, leading to the reduction in water quality and the death of fish and other aquatic life.
- Droughts make wildfires much more likely and we have seen the consequences of that. Runoff carries extra sediment, ash, charcoal, and woody debris to surface waters, killing fish and other aquatic life again by decreasing oxygen levels in the water.
- Drought limits the growing season and creates conditions that encourage insect and disease infestation in certain crops. Low crop yields can result in rising food prices and shortages, potentially leading to malnutrition. You might not feel that will affect you but you might be aware of how quickly your favourite café is raising the cost of a cup of coffee.
- Drought can also affect the health of livestock either directly from the drought or subsequently from crop failure. As with humans, drought reduces their ability to resist infections.
- Dusty, dry conditions caused by drought increase the number of particulates that are suspended in the air,

such as pollen, smoke, and fluorocarbons. Thus asthma and chronic respiratory conditions are worsened.
- Water available for cleaning, sanitation and hygiene reduces or controls many diseases. Personal hygiene, cleaning, hand washing, and washing of fruits and vegetables may be reduced with water shortages resulting in an increased risk of infections.
- Viruses, protozoa and bacteria can pollute both groundwater and surface water when rainfall decreases. People who get their drinking water from private wells may be at higher risk of drought-related infectious disease. Those who drink from treated water are at lower but not an impossible risk. Recreational water use – increased in higher temperatures as we saw – in drought also increases the risk of contracting these infections.
- In periods of limited rainfall, both human and animal behaviour can change in ways that increase the likelihood of other vector-borne diseases. During dry periods, wild animals are more likely to seek water in areas where humans live. These behaviours increase the likelihood of human contact with wildlife, the insects they host, and the diseases they carry.
- Drought reduces the size of water bodies and causes them to become stagnant. This provides additional breeding grounds for certain types of mosquitoes. Outbreaks of Malaria and West Nile virus, which are transmitted to humans via mosquitoes, have occurred under such conditions. Inadequate water supply means people collect rainwater. This leads to collections of stagnant water that become man-made mosquito breeding areas.

Currently about two billion people worldwide don't have access to safe drinking water, and roughly half of the world's population is experiencing severe water scarcity for at least part of the year. Only 0.5% of water on Earth is useable and available freshwater. Over the past twenty years, terrestrial water storage – including soil moisture, snow and ice – has dropped at a rate of 1cm per year, with major ramifications for water security. Water supplies stored in glaciers and snow cover are projected to further decline over the course of the century, thus reducing water availability during warm and dry periods in regions supplied by melt water from major mountain ranges. More than one-sixth of the world's population currently live in areas that get water from these mountains. Sea-level rise is projected to extend salinisation of groundwater, decreasing freshwater availability for humans and ecosystems in coastal areas. Limiting global warming to 1.5°C compared to 2°C would approximately halve the proportion of the world population expected to suffer water scarcity, although there is considerable variability between regions.

Drought numbers will increase with continued climate change and population growth. Droughts are slowly, but surely, heading north and we all should be looking over our shoulders.

Coastal areas

Specific mention should be made of coastal areas and coastal communities. More than 10% of the world's population live in communities that are less than 10 metres (32 feet) above sea level. This is a very large number of people and they are likely to bear the brunt (at least initially/already) of the constant rise in global temperature.

The IPCC figures are clear. Global mean sea level rose by 1.5mm per year during the period 1901-1990, and this accelerated to 3.6mm per year in the period 2005-2022. This acceleration, unless we reduce emissions, will result in sea level rises of at least 1 metre (3 feet) by the end of this century. As well as making all the coastal areas that are at or under 1 metre underwater, the larger tides and more extreme weather events will have a huge effect on all those living on the coast.

We have addressed some of the specific risks to the health of coastal dwellers already but they are summarised below:

- Flooding is obviously the major risk for these areas with all its sequelae.
- Coastal areas are in the path of hurricanes and typhoons as they reach land with maximum power.
- Similarly, coastal communities are primarily affected by tsunamis – themselves more likely as earthquakes and landslide numbers increase.
- Destruction of crops and soils from continued seawater flooding and extreme weather.
- Damage to healthcare infrastructure and responsiveness.
- Rougher seas and migration of fish species to new habitats means further to travel to fish, and in more unpredictable seas.
- Groundwater loss from saline incursion.
- Economic loss from continued structural damage, coastal erosion and loss of livelihood from fishing and dock work.

Globally, sea-level rise will threaten 95% of coastal regions during the 21st century. It is also the most immediate challenge for small island states which could disappear entirely. Universal healthcare is enough of a challenge in developed countries and for those in the developing world even more so. Coastal communities are invariably amongst the poorest in any country with the least resilience and, for many, continuing climate change will not be survivable.

Ice and snow melt

The most obvious and deadly of these is of course snow avalanches. Increasing air temperature and rain under large areas of snow loosens the snow, and it falls as avalanches. Such is the speed and power of avalanches that few things in their path can resist them. Global warming might mean less snow overall on mountains and so apparently reduce the chance of avalanches, but in fact the increased instability of the snow is more significant, hence avalanches will be more common.

Changes in snow and ice integrity also affect the rocks they lie on and bind together. We can expect more rockfalls in mountainous areas and larger rocks to be loosened with obvious consequences. Landslides are already becoming more common both because of increased rainfall but also because of deforestation. Plants and especially tree root systems bind the soil and protect it from being washed away. As we remove more and more trees we make the soil more unstable, hence the increasing number of landslides and mudslides you see on your evening news programmes.

Among the most visible evidence that Earth's climate is warming is the retreat and disappearance of mountain glaciers around the world. Based on data for 2019/2020, 2020 was the 33rd year in a row that glaciers lost rather than gained ice. Glaciers have lost the equivalent of nearly 25 metres (82 feet) of water since 1970 – roughly the same as slicing an average of 27.5 metres (90 feet) off the top of each glacier. Glacier loss means less fresh water in rivers throughout the world – especially in Asia. Destabilisation of glaciers can cause avalanches, landslides and damage dams. As we will see in the next chapter, it is possible that ancient viruses and bacteria have been locked in the ice and are waiting release.

An apparently paradoxical outcome of global warming is larger hailstones. These small frozen blocks of ice that fall from the sky might be thought to be less common in a warming world. The reverse seems to be true with a complex interaction of changes in the atmosphere making them more likely and more destructive. Early in 2023 in Italy, the largest hailstone ever recorded (16cm or 6.2 inches) was measured. Shortly after a hailstone was measured at 20cm (7.6 inches). The latter is bigger (and much heavier) than a tennis ball and if landing on a person could kill and could certainly damage cars and houses. It is a message that bears repetition – we really do not know the full effects of the profound changes we are making to our planet.

Indirect threats to Human Health

The threats to health in the previous section are obvious – especially if you are unfortunate enough to be directly confronted by one of them. In terms of human fatalities though, they are only responsible for a relatively small number (not that each isn't its own individual tragedy). In this section, we will look at the less obvious threats to our health from global warming. These are much more dangerous to us and presently account for many thousands of early deaths, but frighteningly soon this will be hundreds of thousands and then millions.

The cognitive dissonance that we all employ to avoid thinking about the ever-increasing planetary temperature very much plays into the hands of the indirect damage from climate change. If we can't witness something directly we have a way of either not thinking about it or not believing it. As most of us don't suffer failed harvests or starving children or riots over water rights or air quality so poor no-one can go outside, we can pretend they don't represent a threat to us or our children. We see these things on the nightly news ('Don't you find the news just sooo depressing these days') but we can literally and figuratively switch them off.

Unfortunately, your TV remote control can't switch reality off. As more and more of the climate predictions come to pass, at some point our cognitive dissonance will be permanently cured simply by looking out of our own windows.

There are a number of these more insidious risks to our health that we will discuss.

Starvation, malnutrition and food security

We are living in world with increasing atmospheric levels of CO_2. This will have all the effects already described – from higher temperatures to acidic oceans to loss of habitat to animal and plant migration.

But increased atmospheric CO_2 is good for plants – it makes them grow faster as we saw in Chapter 2. This is true and to a certain extent small increases in temperature can increase photosynthesis and allow enhanced plant growth. As we also saw in the chapter, this is not enough to ensure adequate plant growth. Soil conditions need to be right – soil itself is being lost in climate change. Water supply must be right – droughts and floods are destroying crops. Other nutrients must be available – climate change is washing away soil and nutrients, and degrading the bacteria that the plants need to thrive.

The overall effect of climate change is going to be disastrous. Any small agricultural improvement from increased CO_2 will be vastly outweighed by other changes. Many crop plants are very sensitive to environmental temperatures and even a rise of a few degrees will kill them. Water supplies are becoming much less predictable – droughts destroy everything that needs water, floods wash away soils, nutrients and fertilisers. Coastal area agriculture is decimated by higher tides, poisoning of the soil by salt and destruction from wind events. Climate change drives changes in pests, plant diseases and weeds, all of which lower crop yields.

At the extreme end of the scale, climate change directly causes starvation; although the 800 million people who are currently going hungry according to the World Food Programme might not agree about the extreme part of this, as it is happening to them now. Previously fertile land is being destroyed by climate change, deforestation and unsustainable land use. This leads to a process called desertification – literally turning healthy land into deserts. Very little grows in a desert.

Between 1982 and 2015, 6% of the world's drylands underwent this desertification. Anthropogenic climate change has degraded 12.6% (5.43 million km^2) of lands, contributing to desertification and affecting 213 million people, 93% of whom live in developing economies.

It is not just agricultural crops that are suffering, so are the livestock that rely upon them. As well as reduced fodder, these animals are sensitive to rising temperatures – the preferred ambient temperature range for domestic animals is between 10°C (50°F) and 30°C (86°F). Already, even in previously temperate parts of the world animals are dying from heat stress. High temperatures means livestock are more likely to be brought into the coolness of sheds, but this compaction allows greater infective disease spread between the animals and sometimes to their human handlers. Water issues add to this effect, as do wind events and increases in pests or new parasites. A literal perfect storm.

It is important to note the effect ocean warming has on fish and shellfish. Aquatic life forms, as with terrestrial life, are closely adapted to the temperature of their surroundings and if this rises, they will migrate to colder areas. A recent study found that most ocean fish populations were responding to sea warming by relocating towards colder waters nearer the North and South Poles. Vast numbers of people in the world (nearly 10%) rely on the products of oceans and rivers for their nutrition and as sea levels rise, oceans warm and rivers are poisoned, this food supply diminishes. Consequently these populations are facing starvation and malnutrition from protein deficiency.

As the largest traded food commodity in the world, seafood provides sustenance to billions of people. More than 3 billion people in the world rely on wild-caught and farmed seafood as a significant source of animal protein. Freshwater fish are also at risk from warming of the water, increased pollution, increased numbers of new predators, floods and droughts. Fish scarcity not only leads to malnutrition in its various forms, but is a potent cause of conflict. If entire shoals of fish move from your fishing area to someone else's what do you do? Wars are an obvious risk to health but also happen to release astronomical amounts of global warming gases.

Without doubt, food insecurity is increasing and unless we change our current climate trajectory will worsen. As the heat spreads north and south we all need to glance nervously at our supermarket shelves.

Malnutrition. Starvation is the most obvious and severest form of malnutrition. Climate change is among the leading causes of rising global hunger according to the UN's Food and Agricultural Organization. Extreme weather events, land degradation and desertification, water scarcity and rising sea levels, all undermine global efforts to eradicate hunger. Overall, the number of hungry people continues to grow, reaching a total of 821 million worldwide. Starvation kills directly and indirectly – making people more susceptible to other diseases, especially infectious ones.

Other forms of malnutrition are also affected by global warming. As we discussed above, some crop yields may increase with rising CO_2, but unfortunately rising CO_2 levels also affect the level of important nutrients in crops. With elevated CO_2, protein concentrations in grains of wheat, rice and barley, and potato tubers decrease by 10 to 15%. Crops also lose important minerals including calcium, magnesium, phosphorus, iron, and zinc. A 2018 study of rice varieties found that while elevated CO_2 concentrations increased vitamin E, they resulted in decreases in vitamins B1, B2, B5 and B9.

CO_2 levels expected in the second half of the 21st century will reduce the levels of zinc, iron, and protein in wheat, rice, peas, and soybeans. Some two billion people live in countries where they receive more than 60 percent of their zinc or iron from these types of crops. Deficiencies of these nutrients has already caused an estimated loss of 63 million life-years.

There is a myriad of threats to human health from protein and vitamin and mineral deficiencies. Protein deficiency affects over 80 million people worldwide, causing growth retardation, swelling of the legs, liver disease and muscle loss. Liver failure and heart failure are possible if the deficiency is not corrected. Vitamin deficiencies cause a range of disorders ranging from blindness to fractured bones to bleeding and eventually to organ failure. Similarly, a lack of key minerals leads to a range of illnesses and disabilities.

All these deficiencies have significant effects on the individuals and their families and communities. Productive work is lost, therefore exacerbating the nutrition problem. Infectious diseases become rife as malnutrition means resistance to them declines, and so more are then infected and pass that infection on.

Poor nutrition is a direct consequence of climate change from crop failure to loss of livestock, to floods and droughts, to increased insect pests, to the inability to farm in heatwaves.

Insects

From a purely human health point of view, insects present a mixed picture. Without insects there would be no pollination of plants. Without insects dead organic material wouldn't be recycled into the soil. Without insects the soil itself would be detrimentally changed with less turnover and fewer nutrients. Without insects our crop systems would collapse. We need insects for human survival (though the converse does not hold – they do not need us, in fact, quite the opposite). On the other side of the equation, insects damage and destroy crops and livestock. Insects are unwitting carriers of disease to humans and to agriculture.

Insects and infections currently ruin 10–16% of the global harvest. This will inevitably increase as plants are exposed to an increasing number of these pests. Warmer temperatures increase the metabolic rate and number of breeding cycles of insect populations. Insects that previously had only two breeding cycles per year could gain an additional cycle if warm growing seasons extend, causing a population boom. Temperate places and higher latitudes are more likely to experience a dramatic change in insect populations. Some insect species will breed even more rapidly because they are better able to take advantage of such changes in conditions.

So not only will there be an increase in harmful (both to humans and plants) insects, but the insects – driven by climate change – will move to new habitats. These new habitats and the new animals and plants they contain may not have evolved protections against these new predators. Harvests will be lost, crop quality will be reduced, animals will die or reduce production of milk and eggs, thus we enhance even further the risks of human starvation and malnutrition.

Insects also present a more direct threat to human health and we will discuss this in more detail in the next chapter. It is important to repeat that all organisms are closely adapted to their environment. They have evolved to be so and cannot always quickly adapt to change. Sometimes this means extinction, sometimes migration. Non-human organisms don't even know they are migrating, but are drawn to new areas to try to re-establish their optimal temperature, water supply and prey to counteract whatever has changed in their original environment. For humans, this means that new species can present themselves and some of these new species we are ill-prepared for. Insect-borne disease presents one of the greatest threats for us in a warming world.

Microorganisms

Bacteria and fungi will change their behaviour, location and even their biology under the pressure of climate change. Because of their relatively simple structures and rapid reproduction they can adapt to new environments much more quickly than larger organisms. These adaptations will have profound effects upon human health as we are presented with new diseases and lowered ability to treat them.

As we have seen above, agriculture is essential to our species and threats to it are existential. Changes in the bacteria and fungi that cause plant and animal diseases are already occurring and taking their toll on our food supply.

As well as our food being threatened so is our water. To worsen the water supply issues we have discussed previously, we can add the threat of so-called algal blooms. Algae are a group of tiny water-dwelling plants that exist in fresh and cold water all over the world. Their numbers are usually in balance in the water, but changes to that water can cause them to rapidly reproduce and cover huge expanses – this is called an algal bloom. Unfortunately, these algae poison the surrounding water either by producing toxins or dramatically reducing water oxygen. Either of these will wipe out most of the other species in that water. Blooms are promoted by alterations of nutrients from the deeper water, increased temperature of the water and reduced water flow as in droughts – so, all the ingredients that are enhanced by climate change. For those communities who rely on water for food or commerce, these increased blooms will be yet another disaster. The toxins themselves are harmful to humans, and therefore reduce the drinking water availability when they occur. There is also evidence that the toxins they contain can become aerosols, allowing them to travel some distance from the bloom and be breathed directly into our lungs.

Microorganisms do of course threaten our health in a more direct way and are a major cause of disease and death in human populations. They always have and will continue to be so, but it is a frightening scenario that we are creating an environment that makes this threat more potent. Climate change is also predicted to increase the rate of antibiotic resistance. Data from 2013–2015 suggests that an increase of the daily minimum temperature by 10°C (which is conceivable for some parts of the world by the end of the century) will lead to an increase in antibiotic resistance rates of *Escherichia coli*, *Klebsiella pneumoniae* and *Staphylococcus aureus* by 2–4% (up to 10% for certain antibiotics). Human population growth is a related factor and an important factor in contributing to the development of resistance.

Air pollution has also been shown to increase the rate of antibiotic resistance in bacteria. A study published in the *Lancet Planetary Health* found significant correlations between the smallest particulate pollutants and increased antibiotic resistance. This was consistent globally, and in most antibiotic-resistant bacteria and correlations strengthened over time. They estimated that antibiotic resistance derived from this pollution caused nearly half a million premature deaths with over 18 million years of life lost in 2018 worldwide, corresponding to an annual welfare loss of US$395 billion due to premature deaths.

As the climate warms and we get more extreme weather events such as heatwaves, heavy rainfall and droughts, these extreme events create ideal conditions for the spread of infectious disease. Spread of disease occurs more readily via crowded conditions in emergency shelters or the disruption of healthcare services – including delays in routine childhood immunisations. Movement of displaced people brings them into contact with others who may receive or pass on to them infections they have not previously encountered and therefore have no immunity to. Perhaps the most famous example of this was when the Spanish Conquistadores brought non-fatal (for the Spanish) infectious diseases that nearly wiped out the previously unexposed indigenous populations.

I will finish this section by quoting from a paper by Mora *et al* (2022) entitled 'Over half of known human pathogenic diseases can be aggravated by climate change' in *Nature Climate Change*. If we could have any clearer message or a more frightening reason to reverse the disaster we are creating, I don't know what it would be:

> It is relatively well accepted that climate change can affect human pathogenic diseases; however, the full extent of this risk remains poorly quantified. We carried out a systematic search for empirical examples about the impacts of ten climatic hazards sensitive to greenhouse gas emissions on each known human pathogenic disease. We found that 58% (that is, 218 out of 375) of infectious diseases confronted by humanity worldwide have been at some point aggravated by climatic hazards; 16% were at times diminished. Empirical cases revealed 1,006 unique pathways in which climatic hazards, via different transmission types, led to pathogenic diseases. The human pathogenic diseases and transmission pathways aggravated by climatic hazards are too numerous for comprehensive societal adaptations, highlighting the urgent need to work at the source of the problem: reducing greenhouse gas emissions.

Unlike bacteria and fungi, viruses are too simple and small to be directly affected by climate changes such as heat or natural disasters. They are however everywhere and always ready to invade a host. A healthy host can usually fight off a virus but this is not so in those weakened by famine or drought. Insect and animal vectors carry viruses with them, and when the vector numbers increase so does the risk of viral diseases being passed on. When humans are cramped together in small spaces (for example during migration or following a sudden climate event) this produces ideal circumstances for the viruses to infect.

Changes in larger animal behaviour

It is hard to see a corner of our planet, an organism or an ecosystem that will not be profoundly affected by climate change. Even the largest creatures on our planet will be affected and if they are affected, we humans, who crowd them in, will certainly be affected.

As human numbers expand and habitable areas shrink, we are going to come into contact with more of these larger animals. We will increasingly compete for space, food and water. Animals will be forced to migrate further and into areas they have not previously entered in the search for relief from climate change pressures. They are unlikely to obey traffic signs or notice the gardens or farms they trample or devour.

We may not care if elephants, tigers, snakes or sharks become extinct as it doesn't directly affect us. But as we have seen, ecosystems exist in delicate balances and changing this balance can have human consequences that we cannot predict.

The climate and temperature in particular can directly affect animals' behaviour, especially by increasing aggression. Man's best friend is not immune from this, with a study that showed the rates of dogs biting humans increase with increasing temperature and ozone levels. They also observed that higher UV irradiation levels were related to higher rates of dog bites (see the 'Rabies' section in the next chapter). The conclusion was that dogs were more hostile on hot, sunny and smoggy days. To the societal burden of extreme heat and air pollution we can also add the costs of increased animal aggression. Domestic dogs are not fighting us for food or water or space as they already have that, but they are becoming more aggressive as a *direct* result of the heated environment.

Other studies have confirmed this finding of increased aggression caused by raised temperatures and poor air quality in other animal species. From monkeys to salamanders to fish to ants, there is evidence that aggression and conflict increase as the environment stresses these creatures. The competition for resources directly causes an increase in aggressive behaviour and conflicts in animals. A model has been produced that predicts around a 50% increase in animal aggression rates if the warming projected over the next 60 years is as predicted.

To give you nightmares, not only are more sharks heading into previously inhospitable waters as the oceans change, but increased water temperatures have been shown to increase shark aggression. Similarly, studies have shown that increased environmental stress makes snakes more likely to bite. As we increasingly invade their habitats and push them into smaller corners, they will become more aggressive – but heat and drought will also increase this independently. Floods in tropical areas have already created the nightmare scenario of legions of poisonous snakes, scorpions and spiders trying to escape the floods into the same places that humans are heading to. Huge numbers of very stressed poisonous creatures emerging from rising waters was previously the stuff of horror films – but no longer.

Around a million people were estimated to have been killed by animals in 2016 (this excludes humans killing humans). The vast majority of this was via insect-borne infectious diseases, especially malaria, but snakes and scorpions and dogs and hippopotami have played their part. It is very difficult to see any of these numbers declining in the future as we, and the climate, stress them further. This section is about indirect threats to humans but a hungry, charging, heat-crazed elephant or lion will quickly be perceived as a direct threat.

Air quality

Pollution of the air has, to quote the UK government:

> A significant effect on public health, and poor air quality is the largest environmental risk to public health in the UK. In 2010, the Environment Audit Committee considered that the cost of health impacts of air pollution could be as much as £20 billion.

Studies have shown that long-term exposure to air pollution (over years or lifetimes) reduces life expectancy, mainly due to cardiovascular and respiratory diseases and lung cancer. Short-term exposure (over hours or days) to elevated levels of air pollution causes a range of health risks, including effects on lung function, exacerbation of asthma, increases in respiratory and cardiovascular (stroke and heart attacks), hospital admissions and death. Increasing evidence implicates poor air quality in cognitive decline and dementia as well as low birth weight. We will discuss in the next chapter the startling number of medical conditions and diseases that are affected and caused by air pollution.

Poor air quality is a mixture of tiny particles, nitrogen dioxide, sulphur dioxide, ammonia, ozone, carbon monoxide and various other poisons in smaller quantities. This is what you are breathing in, on busy city roads. Your children, who are nearer to the ground, breathe in even more. The source of these toxins is from vehicle exhausts and the burning of fossil fuels in factories and power stations. But these sources are not made worse by climate change (though they are major contributors to it) so why does global warming worsen air quality?

Increased atmospheric heat is more likely to trap the particles and gasses at ground level. Ozone and particulate matter are particularly vulnerable to this and increase considerably on warmer days. As we saw previously, wildfires are far more common on our warming world and release a toxic stew of chemicals often over many miles. Wildfires are becoming more common, more northerly, lasting longer and so releasing more and more pollutants.

The World Meteorological Organization has stated that extreme temperatures are not the only hazard from heatwaves, but that they also cause pollution-related health threats. In their annual air quality and climate bulletin, the meteorologists have highlighted a 'vicious cycle' of climate breakdown and air pollution. They have shown that heatwaves sparked wildfires in the north-western US and heatwaves accompanied by desert dust intrusions did the same across Europe. The hot temperatures in Europe, which in 2022 were record-breaking, led to much higher levels of particulate matter in the air.

We don't need to go into the detailed composition of polluted air. It varies depending on its source but the major health culprits are particulate matter, carbon monoxide, ozone, nitrogen dioxide and sulphur dioxide. Although these are all threats to human health, there is increasing evidence that particulate matter is especially harmful. Particulate matter (also called particle pollution) is a term for a mixture of solid particles and liquid droplets found in the air. Some bigger particles, such as dust, dirt, soot, or smoke, are large or dark enough to be seen with the naked eye. Others are so small they can only be detected using an electron microscope. The sizes of the particles are numerically classified with the bigger particles named PM_{10}; these are the visible particles (soot and dust). $PM_{2.5}$ are much smaller and are not visible to the naked eye. As we will see, these $PM_{2.5}$ are thought to be particularly harmful as their small size allows them to pass through the lungs into the bloodstream where they can affect almost any organ.

An often overlooked pollutant is ammonia, most of which comes from livestock waste management i.e. slurry put on crop fields and fertiliser production. It combines in the atmosphere with sulphates and nitrates to form secondary fine particulate matter, and this being very light can be blown long distances. Ammonia run off from fields can also contribute to the damage to water systems.

Economics, Health and Climate Change

It is easy to ignore the huge effects that climate change will have on the economies of the world. But unfortunately for us all, the health of your country's economy has an effect on the health of its population.

The most comprehensive work on this, *The Economics Of Climate Change,* is also known as the *Stern Review* and is the most definitive review ever undertaken. Its main conclusions are:

- There is still time to avoid the worst impacts of climate change, if we take strong action now.
- Climate change could have very serious impacts on economic growth and development.
- The costs of stabilising the climate are significant but manageable; delay would be dangerous and much more costly.
- Action on climate change is required across all countries, and it need not cap the aspirations for growth of rich or poor countries.

Using the results from formal economic models, the review estimates that if we don't act, the overall costs and risks of climate change will be equivalent to losing at least 5% of global GDP each year, now and forever. If a wider range of risks and impacts are considered, the estimates of damage could rise to 20% of GDP or more. These percentages in monetary terms represent trillions of dollars over the next 10-20 years. This is exemplified by *The Lancet* 2022 report on health and climate change which indicated that from heat exposure alone 470 billion labour hours were lost globally in 2021 – equivalent to nearly 1% of global economic output. This rises to nearly 6% of output in developing economies.

Opponents might say that the expense of mitigating climate change would be just as damaging to our economies. Reducing and eliminating fossil fuels would create huge economic issues for most countries. Investing in renewables isn't cheap and neither is nuclear power. And for either of these to be significant would involve competition for rare earth resources as well as the energy needed to extract them. The point of the

Stern Review is that, however expensive this mitigation will be, it is far cheaper than ignoring the problem. Unfortunately, economics and politics work on too short timescales for significant investment now to be made to protect our future planet.

But what has economics to do with health? Money and possessions do not directly make us less or more healthy. Poorer countries have less money to spend on healthcare such as hospitals, up-to-date equipment, drugs and all levels of practitioners at all levels. Simple public health measures such as clean drinking water, immunisations, sexual health services, health screening and health promotion programmes to name but a few are not feasible without public money. At an individual level, being poor can have a profound effect on your health. If we take a rich, western economy as an example we can see the effect poverty has. The UK British Medical Association's *Health At A Price* reports that 'In England, between 2009 and 2013, the life expectancy for those in the most (economically) deprived areas was reduced by nearly 8 years for men and 6 years for women when compared with less economically deprived areas.' The King's Fund found that, between 1999 and 2010, the majority of areas in England with persistently low life expectancy also had a higher proportion of people earning low or no wages and that the reverse was largely true for wealthier areas.

There are many reasons put forward for these findings. Lower income is associated with a less nutritious diet. Poorer people tend to work more in outdoor jobs with the increased risk of accidents (and of course more exposure to heat, pollution etc.). They are at greater risk of fuel poverty and of being unable to maintain a safe and secure home. They have lower educational attainment – itself an independent risk for reduced lifespan.

Chronic illness is itself expensive – to the individual and to the country. The economic inactivity of the sufferers and the country makes them poorer which is compounded by the increased expense of supporting daily living. As we will see from the next chapter, climate change is going to impact a whole range of chronic diseases. We don't often consider the hidden, economic impacts of these diseases but the evidence shows they can be just as deadly.

In developing and lower income countries, it is obvious that all these problems are magnified. Economics, at all levels, matters enormously. It has direct and indirect effects on all aspects of health. If we continue down the route we are currently travelling and raise the average global temperature by above 2 degrees centigrade, this will, as reported by Stern, have catastrophic effects upon the world economy. These effects will quickly be felt by the population in terms of life expectancy and years of good health – reduced quantity and quality of life for all but a few will be the result.

Politics and Conflict and Migration

Economics inevitably brings us to politics. Politics is affected by climate change and international action is the only way we are going to mitigate climate change. Climate will, despite our politicians' best efforts, slowly creep into every aspect of policy making. As few health threats are not worsened by climate change, so those who govern us will soon realise that no decisions will be made without reference to the changing climate.

We have already seen how economics will be affected by climate change, but there are other major global risks that will be significantly worsened by it. This is already beginning to happen and will define politics for forthcoming generations.

War and conflict have very obvious and direct risks to humans. Equally harmful are the indirect risks of social collapse, destructions of food and water sources and damage to health infrastructure. But does climate change increase the risks of war?

The UN Climate Change Panel recognises the increased risk of armed conflicts as a result of climate change. As in conflicts with non-human animals, flashpoints are related to land, water and food competition. Groups of people who are forced to flee either because of conflict or environmental change impinge upon those already resident, and this causes conflict when resources are limited. There are numerous examples of this worldwide. When the numbers are very large this impingement can lead to all-out war.

Evidence suggests that changes in rainfall patterns amplify existing tensions (IPCC), a prime example being Syria and the role a drier climate played in the country's civil war. However, it would be incorrect to suggest the Syrian conflict was caused by climate change alone. It was but one influencing factor. In water-stressed areas with existing tensions between groups or states over a water source, the impact of climate change on water resources will increase tensions, particularly in the absence of strong institutional capacity.

Sudan's civil war is often described as the first modern climate change induced conflict. The Sahel region is particularly affected due to the region's high dependency on food imports. The global food crisis (mainly caused by the Ukraine-Russia conflict) hit the region hard and further deteriorated food security, especially for the poor. Local prices for rice, wheat, oil, sugar, and other processed imports have already risen between 20 to 50% in different countries in that region. The World Food Programme predicts that in West Africa, seven to ten million additional people could suffer from food insecurity due to the implications of the Sudanese and Ukraine wars.

Countries that have lost agricultural land to climate change may feel forced to attempt to invade a neighbouring country less affected. Often the disputes are at the borders where one country may claim land that they felt was historically theirs. Access to water is and will be an important trigger to conflicts – it may be that risking your life to access a reliable water supply will be worth it.

Conflicts themselves cause damage to agricultural land and poison water supplies, which add to the burdens of the population. The enormous amount of machinery involved in modern warfare emits huge amounts of CO_2. War, through displacement of people and destruction of resources creates its very own vicious feedback cycle. For a modern example of this, in the Kuwait invasion by Iraq during the first Gulf War, 700 of Kuwait's oil fields were set ablaze. The smoke plume above them initially stretched for 800 miles. A staggering 11 million barrels of crude oil poured into the Persian Gulf, creating a slick nine miles long. Inland, nearly 300 oil lakes formed on the surface of the desert, polluting the land for decades. An international coalition of firefighters battled the fires for months until the last well was finally capped in November 1991. Even now, more than 30 years later, the effects of those fires are still felt, with more than 90 per cent of the contaminated soil still exposed.

As governments are rocked by droughts, famines, wars and economic collapse, it inevitably weakens those governments. They have few answers to the problems that face them and this produces weak and ineffective governments. Even worse, to be seen as 'strong', some leaders declare war on a neighbouring country or region. Equally, some might think that hope is lost so ensure that they line their own pockets before it is too late. The only sufferers of this are the people, and the world they live in. Weak or corrupt governments do not do the things necessary to either mitigate or prepare for climate change.

Human Migration

As we have seen previously, climate change is a potent impetus for migration. As environmental degradation accelerates, migration numbers will increase as will the areas they migrate from. Migration is harmful to the health of the migrants, puts pressure on the resources of the areas they move to and reduces the resilience of those they leave behind.

According to the Internal Displacement Monitoring Centre, one displacement (i.e. an individual being forced to leave their home region) **per second** took place due to extreme weather events last year. What is particularly concerning is that given what the science is telling us and what we are seeing in terms of a lack of governmental engagement in preparedness and adaptation, this figure can only increase. The latest

IPCC report affirms the strength of the evidence that climate and weather extremes are increasingly driving displacement across all regions.

Direct health threats to migrants take various forms. One of the most significant is the journey itself. As well as passing through unfamiliar lands that may not have adequate food and water supplies, there are no guarantees that those who already live there will offer assistance and may even exploit them. Migration can increase the threat of drowning through individuals taking transportation risks (such as the highly publicised drowning of migrants in the Mediterranean) or through displaced people encountering unfamiliar drowning risks while on the move or in new locations. Eschbach *et al.* showed that between 1993 and 1997, there were approximately 600 migrant fatalities from drowning in the Rio Grande. It is almost certain that other drowning deaths occur but are not recorded. Most victims are children, women and elderly people crossing the river in small boats or makeshift rafts.

Significant migration can put particular stresses on the area migrated to. Food and water supply may only have been just adequate for the existing population with the new arrivals tipping the balance to starvation and drought. Cramped spaces make infectious diseases more likely as does poor water supply and malnutrition. Cultural clashes may generate conflict between the new and existing populations. Developing countries with weak economies may find the extra humans plunge them into even greater economic trouble. If existing healthcare facilities are limited or stretched, a large influx of struggling refugees may mean complete collapse of that healthcare system.

Significant migration is likely to become a major issue for humans in the coming century and tells us, once again, that for all our sakes, the climate crisis has to be addressed and quickly.

Crime and Climate Change

As we have seen, climate change is so ubiquitous, so all encompassing, that it is hard to find a place or behaviour or system that it will not affect. Plant and animal behaviour is being and will continue to be altered by more hostile environments. We humans feel that we, as the dominant/superior species of the Earth, can control our environment to such an extent that it will not affect our behaviour. The evidence, unfortunately, contradicts this.

There are several ways that climate change promotes criminal activity. As we have seen in the previous section, regions that are most affected by climate change have major disruptions to their economies. Economic hardship is one of the commonest underlying reasons for criminal behaviour. As we also saw, migration presents opportunities for exploitation of desperate people by middlemen, traffickers and opportunists.

Destruction of resources and the resulting shortages promote conflict within and between regions. There is also evidence that climate change and thermal stress independently increase aggression and criminal behaviour.

Peng and Zhan, economists from China, looked at 129 prefectural-level cities in China from 2013 to 2019. They showed that extreme climate events significantly increased crime rates - by 0.035% for every 1% increase in the extreme climate index. They also showed that these had a greater impact on crime rates in eastern areas – which were economically developed and had higher levels of immigration.

Simister and Cooper used USA data to confirm strong seasonal patterns in several types of violent crime. They showed an increase in violent crime (including rape) with increased relative temperatures. There was no fixed temperature for this behaviour to be increased but just a rise in the temperature for that area. Narayan showed that extreme heat events increased incidents of workplace harassment and discrimination for a range of employees.

Anderson summarises the data and highlights various studies and reports that there are about 2.6% more murders and assaults in the United States during the summer than other seasons of the year. Hotter summers produce a bigger increase in violence than cooler summers. Violence rates are higher in hotter years than in cooler years. He also reports that aggression – as measured by assault rates, spontaneous riots, domestic abuse and even batters being hit by pitched baseballs – is higher during hotter days, months, seasons, and years.

We saw previously that animal aggression is increased by heat stress and it seems that we human animals cannot escape this. There are a number of physiological mechanisms put forward for this common behaviour and all come under the umbrella of stress. Animals under very stressful conditions are more likely to exhibit aggressive and antisocial behaviour and heat stress is no different. In humans we can add increased alcohol consumption in hot weather and increased opportunity for crime (from more open windows to more people to bump into outside on hot days). Driving in your own air-conditioned car would seem to be a way to reduce your own personal health risks from climate change. It isn't – car accidents are more common with raised ambient temperatures.

Perhaps paradoxically, the response to climate change can induce its own crimes. We are in a climate situation now where unfortunately significant financial investment by countries and significant individual behaviours will need to change. The money needed to combat climate change will have to be taken from somewhere and this can induce anger and reprisals. For example, areas with a 'polluter pays' policy for car driving have seen significant resistance to this, including criminal damage. When authorities, in an effort to reduce emissions – including improving air quality – reroute traffic this has led to increased vandalism.

These responses to the climate crisis are likely to increase as more and more restrictive measures are forced upon governments. Whilst few would disagree with a cleaner, less polluted planet a significant minority don't want the mechanisms to achieve this to inconvenience them. All those who oppose action now – be they national parliaments or individuals – fail to understand that leaving this issue to the future will result in far more expenditure and far more inconvenience. We really are sitting on the branch we are sawing through.

Other heat induced behavioural changes

We have seen the physiological effects of heat on the body and the causation and exacerbation of diseases. We can add to this increased aggression and crime, but heatwaves disrupt our lives in many other ways – including some surprising ones.

In the next chapter we will deal specifically with the mental health issues associated with climate change but there are more subtle effects. Experiments have shown that increased temperatures reduce our cognitive functioning (perhaps this is also the reason for the increased aggression?) making us less able to do mental tasks. Motor functioning is also reduced with rising heat. These two factors will be particularly significant for outdoor workers and a good employer will realise that protecting their workers from extreme heat is worthwhile.

We are all aware that we are more fatigued in hot temperatures, and this has been shown in laboratory conditions. The Southern European tradition of siestas is, it turns out, a very sensible adaption to the heat. Very warm nights inevitably result in poor sleep and poor sleep is itself responsible for physical and mental illness, which we will again discuss in more detail in the next chapter.

I should add that the response to some of the above issues might be – at least for indoor workers and sleepers – decent air conditioning. That is true, but unfortunately, air conditioning comes at a heavy cost. One study calculated that air conditioning was responsible for the equivalent of 1,950 million tons of carbon dioxide released annually, or 3.94% of global greenhouse gas emissions. Of that figure, 531 million tons comes from energy expended to control the temperature and 599 million tons from removing humidity. The balance of the 1,950 million tons of the carbon dioxide comes from leakage of global-warming-causing refrigerants and from emissions during the manufacturing and transport of the air conditioning equipment. Managing humidity with air conditioners contributes more to climate change than controlling temperature does. The problem is expected to worsen as consumers in more countries – particularly in India, China and Indonesia – rapidly install many more air conditioners.

Finally, to prove beyond doubt that high and prolonged heat changes the behaviour of even the hardiest and most embittered amongst us, is the study by Zong and Zhou. They looked at how Beijing's weather, as measured by the Air Pollution Index (API), temperature and cloudiness (sunny or cloudy), might influence the coverage of the 2008 Beijing Olympics by four US newspapers. The results demonstrated that the API and temperature were significantly related to the negativity of the news reports that were filed from Beijing. Specifically, as Beijing's temperature rose or air pollution level increased, US journalists used more negative words in reporting of the Olympics. The temperature was also correlated with the negativity of China related reports. The findings provided evidence that journalists' news decision-making might be influenced by a greater variety of factors than we previously thought. It is fascinating to think that journalistic reporting can actually be altered by higher temperatures as they write and file their reports.

Overwhelming of healthcare services

We have seen previously in this chapter that many climate induced disasters have a direct effect upon the provision of healthcare. These effects are caused by a number of often overlapping reasons and can overwhelm the ability to respond to emergencies even in the richest and best prepared countries.

Firstly, the enormity of the event might be too great for the healthcare available. This is more likely with sudden cataclysmic events such as extreme wind events or flash flooding as we saw in the 2024 Spanish floods. More prolonged events such as wildfires do allow time for victims to be moved to more distant facilities when local ones are stretched. All this assumes of course that countries already have adequate healthcare infrastructure and the ability to upscale this when needed. Unfortunately, many countries do not have this, and these are also the countries that are likely to suffer the worst extremes of the climate crisis.

Floods, hurricanes and wildfires can also directly impact healthcare facilities. Each has the power to destroy everything in its path – hospitals, clinics and patient transport. Similarly, medical supply depots and storage facilities can be destroyed with loss of medications, dressings, intravenous fluids etc.

Staff usually live in the neighbourhood they work in and will be affected by an event. They may themselves be injured or killed. They may have to deal with family or neighbours rather than coming to work. Transport may be so disrupted that they are unable to attend anyway. Longer term effects on staff including fatigue and post-traumatic stress disorder, may subsequently impact upon the ability of local healthcare services to return to their pre-disaster state.

Disruption of critical infrastructure also presents a significant danger. A hospital or clinic cannot function without an electricity or water supply. If transport links are also disrupted then the necessary technicians to fix these problems cannot access the area. Not all countries have ready access to helicopters to move out severely ill patients or to move in healthcare or infrastructure responders.

Hurricane Katrina, which hit New Orleans in 2005, provides a good example of the problems outlined above – this of course occurred in one of the richest countries in the world and with the highest number of healthcare facilities and staff. Katrina destroyed much of the New Orleans health care system. With more than a dozen hospitals damaged and thousands of doctors dislocated, virtually all New Orleanians lost access to their usual health care providers. Individuals with acute or chronic conditions were particularly hard hit. According to US government officials, 2,500 hospital patients in Orleans Parish alone were evacuated. In addition, dialysis centres across Louisiana with caseloads of between 3,000 and 3,500 patients were destroyed and only half of their patients were accounted for, several weeks after the storm hit.

Critical water and electricity structures were devastated as well as sewage facilities. Road transport was almost impossible for some time after, which limited replacement of supplies and personnel. Katrina was the most expensive natural disaster to date with one estimate at $200 billion dollars. Hospitals were lost and had to be rebuilt. Patients' records were lost and some displaced patients were never seen again. Even 10 years later New Orleans was still trying to recover from the effects. A particularly sobering statistic is that the number of doctors in the New Orleans area declined by almost 75% after Katrina.

Unsurprisingly, Katrina hit the poorest residents hardest. Those who were insured struggled to get healthcare even after the acute event but those without had no chance. This left a trail of ill-health and untreated chronic disease which mirrors those we see in developing countries after these natural disasters. It is very concerning to see the burden that these disasters place upon healthcare systems both in the short term and in the years needed for recovery.

The question of whether healthcare systems are planning for a future of more extreme events, increased infectious diseases, climate induced chronic diseases or climate driven migration is a hard one to answer. That they need to plan for this is obvious and the evidence is already available, but future planning is not a strong point for governments or the institutions they fund. The demand for money in the short term makes it much harder for that money to be deferred to the future. As with COVID-19, it seems likely that we will look back one day and say 'we should have prepared for this'.

Healthcare Costs and Climate Change

Healthcare costs individuals, communities and countries very large sums of money. If climate change worsens global health, then it must increase healthcare costs. The US Natural Resources Defense Council has calculated that the financial costs to our health from fossil-fuel generated air pollution and other climate change consequences are over $820 billion per year.

They analysed this further:

- Soot air pollution: Total annual health costs (2020 dollars) of $820 billion.
- Ozone smog pollution: Total annual health costs of $7.9 billion.
- Allergenic pollens: Allergenic oak pollen was estimated to cause 21,200 asthma visits in the Northeast, Southeast, and Midwest in 2010. Total annual health costs of $11.4 million.
- Climate-fuelled warmer temperatures increase the range of ticks and mosquitos, which carry Lyme disease and West Nile virus, leading to premature deaths,

hundreds of thousands of new cases annually, and tens of thousands of visits to medical clinics and hospitals. Total annual health costs of $860 million-$2.7 billion.
- Heat: Climate change drives higher temperatures and more intense heatwaves, triggering heat stress, heat stroke and worsening a range of cardiovascular ailments, causing deaths and triggering more hospital and emergency room visits. Total annual health costs of $263 million.
- Wildfire smoke: Rising temperatures, drought conditions, and insect outbreaks linked to climate change are projected to increase the frequency and intensity of large wildfires. Wildfire smoke exposure caused 6,200 respiratory hospital visits and 1,700 $PM_{2.5}$ related deaths in a recent year. Total annual health costs of $16 billion.
- Hurricanes: The 2012 Hurricane Sandy disaster caused 273 premature deaths, and more than 12,000 hospital admissions, emergency room visits and outpatient encounters. Total health costs of $3.3 billion.

Once again, those who measure our planet in purely monetary terms often say that fighting climate change is too expensive, that reducing CO_2 admissions would hit everyone's pocket and that net zero (emissions) is financially ruinous. We can see in fact that not dealing with climate change is going to be far more financially ruinous. The figures above are only for one (affluent) country, for one year and do not take into account any other detrimental financial effects of ongoing climate change.

Loss of habitat and drug discovery

Many of the drugs we commonly use were first discovered in nature. Examples would include digoxin, quinine, penicillin, morphine, many cancer treatments and aspirin. As climate change progresses and we lose habitats and ecosystems, potential valuable treatments might be lost forever. What tiny plant or insignificant frog or unclassified tree might contain something of value to us humans?

It has been calculated that the ongoing loss of biodiversity is altering ecosystem functions so much that, in the case of drug discovery, our planet is losing at least one important drug every two years. So even if you care nothing for the natural world, hate trees and rabbits and camels and flowers, and simply feel they exist to serve humanity, you should pause to consider that their loss is more personal than you might hope.

Healthcare and its contribution to climate change

Before ending this section we have to address the contribution of healthcare itself to climate change. Yes, yet another feedback loop and one where global warming puts increased pressure on healthcare which is itself a major producer of CO_2.

The healthcare sector is responsible for almost 5% of global greenhouse gas emissions and has a carbon footprint equivalent to 514 coal-fired power plants. If the sector were a country, it would be the fifth largest polluter on Earth. If we continue as we are, emissions from healthcare will at least triple between now and 2050.

Healthcare contributes to climate change primarily through greenhouse gas emissions. These fall into three main categories. Firstly, emissions which are under direct control of the healthcare facility such as heating and the vast number of vehicles – from ambulances to catering supplies. Secondly, emissions derived from the production of the vast amount of electricity used by healthcare facilities. Thirdly and this is around 70% of emissions, all the various other indirect emissions from the production, transport and disposal of medications, medical devices, hospital equipment and from the movement of staff to and from the facility.

In the global league table of healthcare greenhouse gas emissions, the usual suspects take pride of place. The top three emitters are the United States, China, and collectively the countries of the European Union which comprise more than half the world's total healthcare climate footprint (56%). The top ten healthcare emitters make up 75% of the global health care climate footprint. The United States health sector, the world's number one emitter in both absolute and per capita terms, produces 57 times more emissions per person than does India. While India has the seventh largest absolute health sector climate footprint, it has the lowest health-related emissions per capita. China's health sector produces six times more greenhouse gases per person than India's does; but China's health system also emits one-seventh of the greenhouse gases per capita as does the USA, one third that of Korea, and just under one-half per capita that of the European Union.

This book has not touched upon one of the most obvious and increasingly recognised facts that the main polluting countries are, at least in the short term, not going to suffer the effects of climate change as much as those countries which contribute the least to global emissions. As if more proof were needed, the figures above, which are solely related to healthcare greenhouse gas emissions, illustrate it all too well.

Chapter Summary

- Our warming planet threatens human health in direct and indirect ways.
- Direct threats include wind events, flooding, droughts and earthquakes.
- Indirect threats include malnutrition, changes in animal behaviour and locations, economic issues, conflict, crime and migration.
- The adverse effects of poor air quality are worsened by climate change and are increasingly recognised as a threat to our health.
- Healthcare systems themselves will be affected by climate change so reducing their ability to respond to events.

CHAPTER 5

Disease by Disease

Disease specific threats of a warming planet

We saw in the previous chapter that the migration of some species presents threats to humans in various forms. Migration for those organisms is a form of adaption to what has become a hostile environment. Humans can be very adept at changing their environments (they can literally be on the other side of the world within a day) but in reality still remain vulnerable to significant changes in their surroundings. The environmental change brought on by climate change and the vulnerabilities it brings are something we have already discussed. We have seen how human activity is changing the planet and, if the planet is changing, the risks to our health are inevitably changing.

Increased wealth leads to better health outcomes – having a reliable food source, clean water and a roof over your head are likely to significantly lengthen your lifespan. But this increased wealth has increased the human population and its consumption. More consumption means more energy needs to be harnessed and for the last 200 years most of this energy has been obtained from the burning of fossil fuels. If you have made it this far in the book, you will have realised that this burning has changed and will continue to change our planet. We can't know the full extent of how this will change our environment and we are stepping into a world we have never inhabited before.

What we do know – as we are seeing it already – is that these changes will have an adverse effect on our health. Ironic of course that we have strived to coerce and control our environment to make it safer and more comfortable but have inadvertently threatened our whole existence. We have already been through the major threats to human health, but in this chapter we will see more specifically what diseases and conditions we can look forward to in our warmer world.

Wildfires, volcanoes, insects heading to new climes, invisible air pollution, more aggressive animals, pressure on healthcare facilities from flooding etc. are obviously real but perhaps too vague for us to truly comprehend as a threat. But we do comprehend cancer, liver failure, strokes, malaria or asthma as real threats to us as individuals. We will spend the rest of the chapter discussing these very specific threats. You can read about them all or jump to the ones you are interested in or feel more at risk from. Guaranteed through personal illness or illness of family or colleagues you will have a radar for particular diseases – as we all have. Because they seem so personal, they have a far greater resonance for you than say an announcement of an average 2°C rise in temperature of the planet. If I have one hope for this book, it is that you will realise that you can't separate the personal from the planetary for much longer. The specific diseases we are about to explore, along with the threats described in the previous chapter, explain why, increasingly, it is about **your** health and well-being.

Climate Change and skin disease

We might as well start from the outside of ourselves. This is where we first interface with increased heat or winds or radiation. This is literally the face we present to climate change.

Globally, populations are at increasing risk of skin diseases mainly because of the Earth's diminished ozone layer (despite the Montreal Protocol). The ozone layer is a layer in our atmosphere that acts like sunscreen does on our skin – protecting it from harmful ultraviolet (UV) light from the Sun. The increase in greenhouse gases in the atmosphere depletes this higher ozone layer. This means more UV light reaches the Earth and therefore hits our skin. To add to this, higher temperatures lead to increased UV skin damage at the same dose.

UV light is a potent cause of damage to our skin mainly because it can damage the DNA in our cells directly. The higher the dose of this radiation the greater amount of damage occurs to our skin. Skin cancers are the most serious outcome of increased UV. These are of three types. Melanoma is a cancer of the pigmented cells of the skin. Basal cell carcinoma is the commonest cancer in the world and can cause unsightly sores but is not fatal as it doesn't spread. The third of these, squamous cell carcinoma does spread readily and rapidly. Unsurprisingly, the incidence (the number of people presenting to doctors) of all three is increasing year by year.

Figures show a marked increase in melanoma and non-melanoma skin cancers in the last 30 years. In the USA, basal cell and squamous skin cancers have risen from 400,000-500,000 cases in 1983 to over 5 million in 2012. The incidence of these skin cancers in Scandinavia has more than doubled since 1960, but increased four-fold in Australia over the same period. Even faster growth has occurred with malignant melanomas with a 300% increase in the USA over the last 40 years. Worldwide melanoma has increased between 4 and 5% annually with by far the largest increases in Australia and New Zealand.

Most of the cases of skin cancer occur in fair skinned individuals and especially those who move to sub-tropical countries. Outdoor working or activities (including sun-bathing) also increase the risk. Without doubt, sun exposure is the major risk factor for skin cancer and the reduced ozone protection and increased temperature of our present and future are alarming. Globally over 125,000 deaths are due to skin cancer and the costs to individuals financially, from physical disease and disability and from the psychological burden, are immense. In the US alone it has been estimated that $8.1 billion was spent on skin cancer in 2011.

These are not the only skin diseases that UV exacerbates. There are a large number of inflammatory conditions of the skin (the best known are eczema and psoriasis but there are a myriad more) all of which have one thing in common – they make sufferers lives a misery. These conditions have also been shown to be exacerbated by air pollution, especially from coal burning and from wildfires.

It is not only UV radiation that threatens our skin. Infections from bacteria and fungi are, and will increasingly be, a major health threat in our warming world and skin infections are no different. We will discuss the increase in infectious diseases in more detail later in this chapter, and we have previously discussed how an increase in disease vectors (that is animal carriers of disease) and the warmer air and sea are increasing all infectious diseases.

Studies have already shown that fungal skin infections are not only becoming more common, but are more geographically spread as we warm. It is not just annoying fungal infections such as athletes foot that are increasing, but serious life-threatening infections. One example of this is a fungus called coccidioidomycosis which in around 5% of those it infects can be life-threatening – spreading throughout the body via the bloodstream, happily infecting the lungs, kidneys, joints and brain. It thrives in hot summers and warm winters, and is steadily marching northwards as our temperature increases.

All manner of bacterial infections of the skin also like the warmer weather. A whole host of these bacteria are waiting for the opportunity to infect. Usually a break in the skin is their port of entry, and this is much more likely when other skin diseases such as eczema and psoriasis are present. Once again, the heating of the air and water temperatures allow these bacteria to thrive and await a passing human.

Other nasty infections such as Lyme Disease and Leishmaniasis, which we will discuss in a later section, are also carried by animal and insect vectors and often present with skin manifestations. Both are becoming more common and increasing their geographical range. Being in the water doesn't help you – for example cercarial dermatitis ('swimmers itch'). This delight is an infection caused by avian schistosomes. Schistosomes are tiny parasitic flatworms that live in aquatic snails, hence you get them when in contact with infected water. Warmer temperatures promote the growth of the snails that are an essential part of the schistosomes life cycle. Aquatic birds can also harbour the schistosomes and are more likely to do so in warmer weather where they interface with the snails more often. It has also been shown that warmer temperatures change the migratory patterns of these birds and make them more likely to infect. If further examples were needed of how our interdependency with nature is so crucial, the migration patterns of birds and the effect on a skin disease is a sobering one. When you read of naturalists concerned about climate mediated changes to migration patterns in birds and you think, 'that's awful but nothing to do with me', it turns out you would be wrong.

To finish this section that outlines the environmental war on our skin, we must mention plants. Atmospheric CO_2 enhances the growth of poison ivy, as well as bracken and giant hogweed both of which commonly cause allergic skin disease. All these plants are increasingly being found in regions they have never previously survived in. Something else to look out for on that future nature ramble.

Our skin is in the vanguard of our health in this world of raised temperatures. It suffers some of the worst effects of climate change and is already warning us of worse to come.

Climate Change and lung disease

As our skin is in direct interface with the external environment (and paying the price for this) so are our lungs. The air we draw into our lungs is soon intimately in contact with the cells and tissues of our lungs. In the same way as our skin, our lungs are vulnerable to changes in our atmosphere and environment.

The World Bank describes air pollution and climate change as often being discussed separately, but in fact they are two sides of the same coin. More than this, they both work as a dreadful team that enhance the dangers from each.

Burning of carbon-based sources, be it by fossil fuels from our cars or power stations or from wildfires, will produce carbon dioxide. This CO_2 will rise high into our atmosphere and add to the greenhouse effect. The same burning will release other toxic gases and particles which do not rise into the atmosphere but stay closer to the ground, and these we lump together as air pollution. Poor air quality is a soup of tiny soot particles, nitrogen dioxide, sulphur dioxide, ammonia, ozone, carbon monoxide and various other poisons in smaller quantities.

As we discussed in the previous chapter, global warming, by increasing the trapping of these particles and gasses at ground level, worsens air quality. Ozone and particulate matter are especially affected by this, and so increase considerably on warmer days. It is also important to remember that, in our heating world, wildfires are becoming more common, more northerly, lasting longer and so releasing more and more pollutants. Note that Ozone in the upper atmosphere, though a greenhouse gas, does protect us from UV light but when produced by chemical reactions at ground level is a major pollutant.

Air pollution is now the world's leading environmental cause of illness and death. Fine air pollution particles or aerosols are responsible for 6.4 million deaths every year. It has been estimated that the cost of the health damage caused by air pollution amounts to $8.1 trillion a year, equivalent to 6.1% of global GDP.

Each day we are breathing in this toxic group of chemicals and no matter where we live these toxins are increasing. Those in urban environments near roads or industrial areas are exposed to the highest levels. In developing countries, most of the air pollution is domestic from burning of fuels for cooking or heating. Whatever the source, the higher the levels of pollutants the higher the level of ill-health.

Sandstorms are also particularly hazardous to our lungs as it is difficult not to breathe in these larger particles. Sandstorms are caused by a combination of heat, wind, and loose sand or soil. The latter are more likely in areas where plants that should bind the soil have died off following droughts. Climate change is a driver for all these factors.

It is obvious from the above that one of the main deleterious health effects of global warming will be upon our lungs. We take in about 15 breaths per minute of air and this air passes immediately to our lungs and is then absorbed into our circulation. As well as the food we eat, we are the air we breathe.

Asthma. Asthma is a very common lung disease and can be thought of as a narrowing and blockage of the fine tubes in the lungs that carry the air. The narrower these tubes are the less air can move in and out of the lungs so less oxygen gets into the bloodstream. Asthma can vary from mild to life-threatening but either way has a significant cost to individuals and to healthcare.

Pollutants and irritants in the air are potent triggers for asthma. These trigger the airways in the lungs to narrow, thus causing the shortness of breath and wheezing of asthma. The higher the air concentration of these irritants, the more likely an asthma attack will happen and potentially the worse it will be. An asthma attack is a frightening experience for the sufferer and witnesses as the person literally fights for breath.

There is another aspect of climate change which significantly exacerbates lung diseases and especially asthma. Allergens are substances that are seen by the body as foreign and so produce a protective reaction by the body in an attempt to defend against this foreign invasion. This is a natural and important process but in some individuals there is too great a response which in itself becomes damaging – this is called an allergic reaction. These can occur in different parts of the body with varying effects, but in the lungs they trigger an asthmatic response (i.e. the airways narrow). This is a different mechanism (though related) from the direct toxicity of emissions. The most common outdoor allergens are pollen and fungal spores.

There is evidence that climate change has increased the number of allergens in our air. Warmer atmospheric temperatures have allowed earlier season plant growth so earlier pollen release and longer pollen seasons. There is also evidence that raised CO_2 levels have independently increased the level of pollen produced by plants. Remarkably, there is now some evidence that some of the pollens produced are **more** allergenic because of raised atmospheric CO_2 levels. As if more threats were needed, there is also evidence that airborne pollutants can exacerbate the allergic response to pollen and fungal spores.

Significant increases in allergic diseases are already being seen worldwide – especially asthma and hay-fever. Models predict a steady increase in sufferers if we continue to pollute. We can see from nightly news items wildfires somewhere in the world, we continue to increase CO_2 levels in the atmosphere and clean air zones in cities are constantly undermined by vested interests. We will create more children and adults with asthma and asthma attacks will worsen with increased mortality and healthcare costs. Does it take us to be literally gasping for air to convince us to change our ways?

<u>Chronic Obstructive Airways Disease (COPD)</u>. This is really an umbrella term for diseases that result in inflammation and damage to the lungs which in turn cause blockage of the airways, infections and breathlessness. There is also usually an element of narrowing of the small airways as with asthma. COPD affects 12-16 million people in the USA alone and is the third leading cause of death. The World Health Organisations estimates that air pollution is responsible for over 3.7 million premature deaths worldwide, 14% of these due to COPD.

As it is a chronic disease it has a serious effect upon quality of life with sufferers having chronic cough, wheeziness, intermittent infections, reduced exercise tolerance, poor sleep, reduced employment prospects as well as shorter life spans. It is expensive for the individual with reduced income and is a major burden to healthcare systems.

The vast majority of COPD is caused by environmental factors, with the main culprit being that most personal of all air pollutants – cigarette smoking. Increasingly though, it is becoming clear that as smoking numbers decline, outdoor air pollution is playing a greater role in the disease. Particulate matter has been shown to be a potent cause of COPD, as have the polluting gases nitrogen dioxide and ozone. Additionally, increased atmospheric temperature has been shown to exacerbate COPD significantly.

Once again we can see that the climate we are already creating – higher temperatures, increased emissions in the upper and lower atmosphere, more intense wildfires and increasing urbanisation create a perfect storm of factors to reduce the quality and quantity of human life.

<u>Lung Cancer</u>. Lung cancer is a feared and often deadly cancer that has seen its prevalence (i.e. how many people have the disease) dramatically rise in the last century. This has of course been definitively shown to be caused by smoking. However, there is also strong evidence that outdoor air pollution is linked to lung cancer and the International Agency for Research on Cancer has classified outdoor air pollution as carcinogenic to humans.

The European Society for Medical Oncology reported in 2022 that:

> The same particles in the air that derive from the combustion of fossil fuels, exacerbating climate change, are directly impacting human health via an important and previously overlooked cancer-causing mechanism in lung cells. The risk of lung cancer from air pollution is lower than from smoking, but we have no control over what we all breathe. Globally, more people are exposed to unsafe levels of air pollution than to toxic chemicals in cigarette smoke, and these new data link the importance of addressing climate health to improving human health.

We are now seeing, and will continue to see, more and more lung cancer in non-smokers. This type of lung cancer is called non-small cell lung cancer and currently accounts for over 250,000 lung cancer deaths globally per year. Whilst with the reduction in the number of smokers (at least in Europe and the USA) lung cancer rates were thought to be diminishing, we can now see this will not be so.

COVID-19 and climate change. COVID-19 needs little introduction, and there are other similar recent viral epidemics but with less global fame. COVID-19 was caused by a virus called coronavirus. As yet there is no evidence that, unlike bacteria and fungi, global temperature rise **directly** affects viruses. However, there is evidence that, as the planet heats up, we will become more susceptible to viral infections. Be it malnutrition causing us to be more susceptible to infection, or flood water carrying dangerous viruses, there will be many ways climate change will result in greater disease and death from viruses.

One increasingly important risk comes from the wild animals that can be the source of these viruses. As they lose their natural habitats to agriculture, wild animals both big and small and on land and in the sea, are increasingly sharing a smaller space with the increasing number of humans. This closer contact means that viral diseases can more easily jump from animals to humans (the cause of the recent epidemics). Related to this, the increased human demand for meat involves larger livestock farms which again increase the risk of viral diseases jumping to our species.

Developing Lungs. It has been shown that exposure to air pollution in early life can have a long-lasting effect on lung function. There is evidence that the process of normal lung function growth in children is suppressed by long-term exposure to air pollution. This means asthma is more likely, and more likely to be severe. Throughout childhood, there is a natural development of lung function and maximising this is important as poor lung function leads to less reserve in adulthood if lung disease develops.

Climate Change and heart and circulatory disease

Cardiovascular disease is a general term describing conditions that affect the heart and blood vessels, and therefore the circulation of blood through the body. Like water going through the pipes in your house, a smooth flow through healthy pipes (blood vessels) and a healthy pump (the heart) are essential. Damage to the pump or narrowing of the vessels means a reduced blood supply and our tissues will die without an adequate blood supply.

If the blood supply is suddenly reduced to the heart this is what we call a heart attack. If it is suddenly reduced to the brain then this is a stroke. We can have blockages to the blood supply of any part of the body from the legs and arms to the eyes. Disrupted blood supply also causes clots to form, and if big enough this might present as a lung embolus. There are many other manifestations of cardiovascular disease and many cause serious disability or death. Even those with less dramatic consequences are often heralds for more serious disease in the future.

Already, according to the World Heart Foundation, deaths from cardiovascular disease (CVD) jumped globally from 12.1 million in 1990 to 20.5 million in 2021. It was the leading cause of death worldwide in 2021, with four in five CVD deaths occurring in low and middle-income countries. It is also a major cause of disability worldwide with an estimated doubling of those living with cardiovascular disabilities from 17.7 million in 1990 to 34.4 million in 2019.

Does climate change increase the risk of heart attacks and strokes? By now, you won't be surprised to know that it does. There are four main factors that increase this risk.

Heat. Chronic exposure to increased heat has been shown to increase cardiovascular events in populations – especially those who do not have access to mechanisms to cool such as air conditioning, access to rivers or lakes or transport. Acute exposure to heat in the form of heatwaves has long been known to trigger heart attacks and strokes. This is especially so with high humidity (see wet-bulb temperature in Chapter 3). In an attempt to cool down, blood is diverted to the skin so that heat can be dissipated into the air. This means that the heart has to work much harder to pump blood around the rest of the body and this increased work by the heart is what can precipitate a heart attack.

In very hot weather, the blood also becomes stickier and so more likely to clot. This is again related to the blood being diverted to the skin as well as dehydration. Blood that is more likely to clot makes heart attacks more likely as well as strokes, deep vein thrombosis (DVT) and lung embolus. In young, healthy individuals these cardiovascular events are unlikely to occur, however if heatstroke develops (see Chapter 3) then no matter how healthy you are the risk of life-threatening events dramatically rises (so don't go running in a heatwave).

Liu *et al.* have attempted to quantify these heat related risks. Their meta-analysis (i.e. a synthesis of all previous studies) of 266 papers indicated that a 1°C rise in environmental temperature was associated with a 2.1% increase in cardiovascular diseases related mortality and a 0.5% increase in cardiovascular disease related morbidity (which means significant disease or disability but not death). The strongest effect was an increase in strokes by 3.8% and heart attacks by 2.8%. The risk of cardiac arrest (that is the heart stopping completely – which is not the same as a heart attack but may be caused by one) was as high as 2.1%.

For heatwaves, cardiovascular disease related mortality increased by a staggering 11.7%, and the higher and longer the heatwave the higher this figure is. Heatstroke is life-threatening, with an early mortality of over 50%. By the 2050s, heat stroke-related deaths are expected to rise by nearly 2.5 times the current annual baseline. Heat kills and in an ever warming world, that information by itself, should make us panic.

Air pollution. We saw in the previous section the all-pervading dangers of air pollution. As well as its effects on the lungs, it has measurable adverse effects on the heart and circulation. The fine particulate matter ($PM_{2.5}$) has been shown to increase inflammation in the blood vessels and to create harmful chemicals in the bloodstream. The inflammation damages the blood vessels, so clots are more likely to form. The blood itself becomes more viscous making it more likely to clot. The changes also increase blood pressure over the longer term. All these combine to dangerously increase the risks of heart attacks and strokes.

The risk of heart failure (the heart failing as a pump), heart attacks, abnormal rhythms of the heart and stroke are increased by both short and long-term exposure to air pollution, especially in susceptible individuals. This includes older people and individuals with pre-existing cardiovascular and respiratory conditions.

Vector-borne diseases. We will be discussing infectious diseases and the animals that pass some of these infections on to us later. A number of these infections can cause direct damage to the heart. Lyme disease can cause heart inflammation and damage. Chagas disease (which affects 8-12 million people in South America but is creeping north into the USA) can have all sorts of serious effects on the heart. Dengue fever (also creeping northwards into Europe) can cause inflammation and damage to the heart and blood vessels.

<u>Mental health.</u> High heat and humidity have been shown to have an association with substance use disorders, schizophrenia, mood disorders and anxiety. All these conditions are themselves associated with increased cardiovascular disease. We will discuss the effect of climate change on the wider aspects of mental health later.

Added to all this, behavioural changes that heat induces can have an effect upon disease incidence. Higher temperatures reduce outdoor activity. People are less likely to go for a run, to exercise, to do team sports. Inevitably they are more sedentary, and this significantly increases their risk of heart disease, diabetes, obesity and cancer. It also means more cooking is done inside and often with inefficient ovens or fires – themselves a risk for the production of toxic particles and gases.

Climate Change and cancer

We have already seen the increased risk of skin and lung cancer with continued climate change and warming. Is there is an increased risk of other cancers, that most feared disease, in other organs and sites resulting from the same environmental changes? Unfortunately, for us all, there is.

Cancer is widely predicted to be the leading cause of death in the 21st century, overtaking heart disease. Worldwide, there were 24.5 million new cases of cancer and 9.6 million deaths in 2017, a striking increase from 2008 with 12.7 million cases and 7.6 million deaths. Though there are many factors underpinning this there is certainly, at the very least, circumstantial evidence that climate change is a contributing factor. The harder evidence is presented below.

Air pollution and climate change are intimately linked and air pollution, as we have seen, is a potent cause of lung and heart disease, including lung cancer. It has also been associated with other cancers including mouth, oesophagus (gullet), stomach, liver, bowel and breast. As we discussed in the previous section, significantly increased outside temperatures prevent people going out, both reducing their activity and increasing their risk of exposure to toxins from cooking. Indoor emissions from coal are known to contain some high-risk carcinogens.

Yet again, wildfires show how damaging their emissions can be. A Canadian study found those exposed to a wildfire within 50 km of residential locations in the past 10 years had a 5% relatively higher incidence of lung cancer than unexposed populations and a 10% relatively higher incidence of brain tumours.

It has also been predicted that changes to our climate and the resulting issues with agriculture and food supply will have a detrimental effect. It has been estimated that 30–40 percent of all cancers can be prevented by lifestyle and dietary measures alone. Protective nutritional elements in a cancer prevention diet include selenium, folic acid, vitamin B-12, vitamin D, chlorophyll, and antioxidants such as the carotenoids (α-carotene, β-carotene, lycopene, lutein, cryptoxanthin). With the reduction in crop yields in our warmer future, it will become harder to obtain the fresh fruit and vegetables required for this. Additionally, as we saw in Chapter 4, the nutritional value of each fruit and vegetable is reduced when the plants are stressed by greater heat.

More specifically, a diet containing fresh plant derived food with the correct levels of vitamin and minerals has been shown to give some protection against breast, colorectal, and prostate cancers and even lung cancer. We can extrapolate that these are the cancers that will be increased the most as the world's plants wilt under the increasing temperatures.

We have already seen that patterns of infections will change as the world does. This will increase infections and some of these infections can result in cancer. For example, the parasitic fluke (worm) *Schistosoma hematobium* found in Africa and the Middle East causes bladder cancer. *Opisthorchis viverrini* is found in South East Asia and causes a particular type of liver cancer. Both will thrive in a warmer world and there may well be other infective agents awaiting that we are not even aware of yet.

In the previous chapter we discussed the importance of the disruption of healthcare systems from severe weather. As well as direct damage to buildings, supply lines are disrupted and staff affected. This can limit the response to severe climate events, but once these events are over the strains on healthcare systems remain. Cancer screening, cancer diagnosis, cancer treatment and cancer follow-up all need well-functioning healthcare systems. Extreme weather events will disrupt this both from the healthcare providers' point of view and the patients. In developed countries this will delay treatments, but there should be the capacity to catch up or to have neighbouring institutions help out. In less developed healthcare systems this extra capacity may not be available.

In those countries without resilience, ongoing cancer care or screening is likely to be affected – perhaps permanently for some. Even in developed countries, if the extreme weather events keep on coming then they may find it increasingly hard to maintain their core services – including for cancer. Already climate change is having a serious effect on the economies of countries all over the world. Healthy economies allow more to be spent on the health of the populations and the converse is obviously true.

In our battle against cancer, climate change wouldn't at first seem to be an obvious or particular threat, but when we look more closely it is. Just as we might change our own behaviour purposely to try to reduce our cancer risk, we should remember that we need to protect our planet to protect ourselves. We should see the opportunity in this as improving our own health and the health of our planet can often go hand in hand (a simple example might be to walk to your destination rather than drive).

Incidentally, if you smoke then you have a vastly increased risk of dying from cancer or heart disease. But more than likely you know that. What you probably don't know is the heavy toll the tobacco industry has on the planet. Vineis *et al.* report that just one tobacco manufacturer, in one year, emits 4.5 million tons of CO_2. They also use 23,247 thousand cubic metres of water per year. It's an industry that pollutes the individual and the planet.

Climate Change and kidney disease

You may not think of your kidneys very much, and certainly not in relation to climate change. However, your kidneys have many functions but the most important is they control our fluid balance. They therefore play a vital role in protecting us from our ever-warming environment and consequently are directly threatened by the heat.

The main function of the kidney is to ensure that the fluid and salt balance of our bodies is maintained within narrow limits. It does this by sensing concentrations of salts in our blood and taking the appropriate action. Too dilute and we urinate, too strong and we stop passing water. Thus, on a warm day the kidneys are working hard to maintain our fluid balance and indirectly therefore our core temperature. Kidney failure – acute or chronic – has very serious detrimental effects on our ability to regulate fluid and therefore heat.

Heatstroke, which occurs when the core body temperature rises (hyperthermia), as we have previously seen is a life-threatening situation. Recovery from the acute event does not necessarily result in a return to normal health. Acute kidney injury is a common manifestation of heatstroke – in the 1995 heatwave in Chicago over 50% of those presenting with heatstroke had acute kidney damage. These acute cases showed significant changes in blood salts such as Sodium, Potassium and Phosphate. These changes in salts can directly damage tissues and if levels are too far out of the normal range are fatal.

Around 10-30% of those who suffer heatstroke with associated acute kidney injury require dialysis (even the most well equipped dialysis units will be stretched by these numbers). Most who survive the acute event – even those who needed dialysis – have a kidney function that returns to normal but a portion progress to chronic kidney disease requiring life-long dialysis or transplant.

Occupationally induced/enhanced heat stress seems to be a particular risk for the kidneys. Epidemics of chronic kidney disease have been identified in various hot regions of the world in manual labourers working outside. Sugarcane workers in Central America have been studied quite extensively, and this increase in kidney disease seems to be the result of repeated episodes of heatstroke with resulting and accumulating kidney damage. Similar patterns of kidney function problems have been identified in India, Sri Lanka, Mexico and even Florida and California. As the temperatures and humidity rise even those who have less demanding physical work will begin to be affected – at risk individuals, employers and governments take note.

Heat stress and dehydration make the urine more concentrated and when this happens there is a risk of kidney stones. These can be of various types but all are caused by urinary salt imbalance allowing these salts to become solid instead of staying in liquid. They are very painful when they pass through the urinary system but if they don't pass through they can block this system and cause kidney damage. The 'Kidney Stone Belt' refers to the region in the south eastern United States where the rate of kidney stones, or kidney calculi, is excessive. North Carolina reportedly has the highest incidence of kidney stones in the nation though the reasons are not clear. What is clear is that it is projected to move to other states as our climate warms further.

Similarly, concentrated relatively static urine is more likely to become infected. This is both discomforting and alarming but usually not serious. However, repeated bouts can cause permanent kidney and bladder damage – which can in itself lead to a higher risk of further infections.

As we have previously discussed, the pressure these extreme events put on healthcare systems should not be ignored. Too great a pressure may overwhelm facilities – in the acute and the recovery phase. One study from New York in 2022 found that extreme heat exposure days were associated with a 1.7% to 3.1% higher risk of emergency department visits related to kidney disease. The association was stronger with a greater number of extreme heat exposure days and the effect lasted for an entire week following the heatwave. The association was related to acute kidney injury, kidney stones, and urinary tract infections.

Chronic kidney disease (CKD) already consumes a large amount of the world's healthcare resources. One analysis reported that in 2017 the global prevalence of CKD was 9.1% (697.5 million cases). Nearly one-third of all cases of CKD were in China (132.3 million) or India (115.1 million), 10 countries had greater than 10 million cases and 79 countries had greater than 1 million cases. The all-age and age-standardised global incidence of dialysis and kidney transplantation also increased between 1990 and 2017 (by 43.1% and 10.7%, respectively, for dialysis and 34.4% and 12.8%, respectively, for transplantation). CKD resulted in 1.2 million deaths and was the 12th leading cause of death worldwide. In addition, 7.6% of all CVD deaths (1.4 million) could be attributed to impaired kidney function. Together, deaths due to CKD or to CKD-attributable CVD accounted for 4.6% of all-cause mortality. Global all-age CKD mortality increased by 41.5% between 1990 and 2017.

The rise in the number of cases is mainly a result of our ageing population, but it is obvious that if spiralling global temperatures put even more people into kidney failure, healthcare systems worldwide are going to have to deal with a significant extra burden.

Finally, for your own protection it is worth remembering, if you treasure your kidneys, avoid excessive heat, do not exercise in high temperatures, if working outside take as many precautions as you can and hydrate well. This latter point needs a note of caution added – firstly don't overhydrate as this also puts too great a strain on the kidneys, take regular small sips of fluid and use your thirst (that's what it is there for) to guide on any extra intake. Avoid rehydrating with fructose containing drinks as they have been shown independently to damage kidneys after exercise.

Climate Change and liver disease

The liver has been simplistically described as the factory of the body. It certainly handles a huge range of functions and without a liver there would be no you. Because of its high workload it has a very large blood supply, and it cleans and detoxifies this blood that passes through it. Everything that we eat is absorbed by the bowel, and then passed to the liver to be dealt with and distributed to the correct tissues. This also means that any toxins we ingest are immediately passed to the liver, which has to try to deal with them to prevent them harming us. Unfortunately, these toxins, if in sufficient quantities, will damage the liver itself.

Liver disease causes approximately two million deaths per year globally, and this burden of liver disease continues to grow. A US study reported that from 2012 to 2016, the rate of chronic liver disease related hospitalisations per 100,000 hospitalisations increased from 3056 (95% CI, 3042-3069) to 3757 (95% CI, 3742-3772). Total inpatient hospitalisation costs increased from $14.9 billion to $18.8 billion. An increase in liver disease already presents an alarming future healthcare burden but will climate change make this even worse?

There are a range of factors as our climate changes that will directly impact upon liver disease. As the liver has a diverse, but pivotal range of functions the list of potential damage is diverse.

Particulate pollution – as with many of the diseases we have already discussed, particulate matter in the blood stream damages the liver tissue directly. Long term exposure to particulate matter has been shown to cause abnormalities in liver enzymes. Liver enzymes in the blood are a very useful marker of liver problems as they show early and increase proportionately with greater damage.

The particulate matter can also carry other toxins into the body which again damage the liver. This has been shown for toxic fungal spores, heavy metals and especially for those living near to petrochemical plants.

Polycyclic Aromatic Hydrocarbons – these compounds are produced from incomplete burning of organic matter and come from waste burning, factories, wildfires and inefficient heating and cooking systems. There are a number of different polycyclic aromatic hydrocarbons but most are recognised as toxic and long term exposure causes liver damage. As we now know, wildfires are far more likely to occur in our warmer world and are an increasingly common source of these poisonous substances.

Heavy metals in water – lead, chromium, arsenic, mercury, nickel and cadmium are toxic metals and are particularly toxic to the liver. They cause abnormal liver function, scarring of the liver, cirrhosis and sometimes liver cancer. Though present naturally in the soil and water in small amounts, they become concentrated in the vicinity of industries that use them – either as inadvertent or purposeful leakage into the land. As flooding increases with climate change, more of these are washed into rivers and streams i.e. drinking water. During droughts, as the water level drops these metals become more concentred in the water. Direct toxicity may occur to humans who drink this and indirectly they may eat meat or vegetables contaminated by the metals in this water.

Infectious diseases – a large number of blood-borne infectious diseases affect the liver. This is hardly surprising; with a large blood supply and well-nourished tissue it is an ideal home for parasites. Its close connection to the gut means that water and food-borne infective agents quickly meet the liver. We know from previously that these infections – bacterial, fungal, parasites – are already increasing in number and range due to rising temperatures and migration of insect hosts. These infective agents are going to be responsible for a significant number of liver problems in the near future. We will discuss the commonest liver infections in the infectious diseases section, but a flavour of the risk is given in a paper by Donnelly *et al.* in *The Journal of Hepatology*:

Increasing ambient temperatures are changing the geographical distribution of hepatic (liver) parasitic infections, including schistosomiasis. There have been unexpected outbreaks of schistosomiasis among swimmers in the Cavu River in Corsica, for example. Climate change is reported to have contributed to increased *Fasciola hepatica* (liver fluke) in the United Kingdom, and it is estimated that there may be epidemics of infection in Wales by 2050.

<u>Malnutrition</u> – this is often wrongly thought of as only meaning not having enough to eat – but it is more than that. Those who have enough to eat but are not getting adequate amounts of vitamins and minerals are also malnourished. Those who overeat ultra-processed food and become obese are also malnourished. We will deal with the link between obesity and climate change later in this chapter, but increasing rates of obesity worldwide are fuelling the rise in liver disease. Those who do not have enough to eat, especially lack of protein, develop liver disorders. Adequate intake of vitamins and minerals is as important for the liver as any other organ, and its high work rate means that it is affected early when deficiencies occur. This can be compounded as liver failure results in poorer storage of these essential nutrients, resulting in even lower levels in the body.

<u>Blue Green Algae</u> – we saw in Chapter 4 that as a result of global warming there are increased incidences of contamination of drinking water with poisonous algae. Cyanobacteria, which is also known as blue-green algae, can invade and choke stagnant water supplies. These algae produce toxins and when taken in drinking water they poison the liver cells causing liver inflammation and progressive liver damage. A study has shown that higher amounts of this algae, as a result of adverse weather

conditions, may have contributed to an increase in liver disease mortality in the United States.

Heatstroke – unsurprisingly heatstroke can have a deleterious effect upon the liver. Raised core temperature has been shown to directly damage the liver. The severe reduction in central blood supply – as so much has been diverted to the skin to reduce temperature - has a particularly profound affect upon the liver. The reduction in blood and the rise in core temperature causes liver damage; the liver can't respond adequately to this damage and can't repair itself because of an inadequate blood supply and so further damage occurs. Heatstroke is therefore directly related to both acute liver injury and sometimes to complete liver failure.

Alcohol and Mental Health – communities, countries, continents that are increasingly stressed by climate change suffer. They suffer from economic strain, from overstretching of healthcare facilities, from agricultural failure, from inadequate water supplies. Individuals are stressed in the same way and this takes its toll on their mental health. A not uncommon way of dealing with this stress is substance abuse – especially alcohol. Few of us are unaware of the effects of alcohol on our liver – alcohol use was associated with 3 million deaths and 132.6 million disability-adjusted life years (DALY) in 2016 alone. Of those aged 15–49 years, 12% of male deaths and 4% of female deaths were associated with alcohol-related diseases and injuries in 2016 worldwide. As the pressures from our climate increase, it is likely that alcohol consumption (and other substances such as cocaine and marijuana, both of which have been linked with liver damage) will increase and the results are already obvious.

Climate Change and gut health

We can compare the lining of our gut to our skin and lungs in that it comes into direct contact with the outside world. All our nutrients and water are absorbed by it – beginning in our mouths and finishing in our large intestine. Any toxins or infective agents that we might also inadvertently swallow must pass into our gut. Our gut does protect us from these with its own defences to some extent but equally can be seriously harmed by them. Climate change, as we have repeatedly seen, is changing and will continue to change our environment, and this inevitably will affect our gut environment and the important bacteria that live in it.

Perhaps the most likely effect of changes to the gut from climate change will be changes to water quality allowing diarrhoeal-causing organisms to flourish. Rising air and water temperatures and flooding will lead to an increase in water-borne infectious diseases – one notorious example of this being cholera. Cholera is caused by a bacteria that lives in water and is present when raw sewage flows into rivers and streams. When this infected water is drunk, the bacteria fastens itself to the gut wall preventing water and nutrients being absorbed, hence the diarrhoea. If untreated the sufferer will quickly die of dehydration.

Cholera is just one example of the gut infections that cause much misery and death – there are a number of others such as salmonella and campylobacter as well as the blue-green algae we discussed previously. Diseases from these and other pathogens will get more and more common as we warm. But we don't need to wait for evidence for this as the incidence of cholera is already rising. The WHO reported in 2023 that since 2021 the number of cholera outbreaks reported has increased worldwide; more and more countries are reporting outbreaks and the fatality rates are rising. Overall, diarrheal diseases are one of the leading causes of death in children. The World Health Organization estimates that by 2050, about 33,000 extra children under the age of 15 could die from them worldwide due to climate change.

Cholera represents another herald of what is to come if global temperatures keep rising. Warmer water encourages the cholera bacteria to thrive, floods disrupt sewage systems, droughts force people to drink infected water, economic pressures on governments make them less able to protect water supplies so allowing conditions for cholera to flourish. Those in the West may think of this as a disease of Africa and Asia, but cholera has been with us before and can come for us again.

Less obviously, but potentially of equal seriousness, is the change in our normal gut bacteria. We have a very close relationship with the bacteria that happily live in our bowels. These are not disease-causing bacteria, but live with us in a symbiotic relationship i.e. both us and the bacteria gain from this. It is a complex relationship which we won't go into, but these bacteria are essential for our continued health. It has been shown that microbial communities present in the gastrointestinal tract (gut microbiota) are sensitive to changing climate, especially temperature and humidity. Thus we know already that gut bacteria are altered both directly and indirectly by climate change but what we don't know is the long term effects this will have on our health.

Studies have shown that global warming can alter the bacterial ecosystem of the soil. This can then change the soil cycling of carbon, phosphorus and nitrogen – all essential for healthy soil and plant growth. The loss of soil biodiversity due to a reduction in the number of bacteria can subsequently deplete human gut bacteria. Lower organic content in our diet also impairs the normal functioning and metabolism of gut microbiota and this results in nutritional problems for us host humans.

There is some evidence – though not as strong as with respiratory and cardiovascular diseases – that gut health may be affected by air pollution. The so-called Inflammatory Bowel Diseases – Crohn's and Ulcerative Colitis – appear to be at an increased risk in high pollution areas. Kaplan *et al.* using data from the UK, found a possible link with vehicle pollution with individuals under 23 years being more likely to be diagnosed with Crohn's disease if they lived in regions with high Nitrous Oxide levels. Ulcerative colitis patients under 25 years were found to be more likely to live in regions of higher Sulphur Dioxide.

Gastroesophageal reflux (heartburn) has also been linked to air pollution levels. A Korean study found an increase of medical treatment for reflux in areas with high pollution levels. They found a specific association with raised particulate matter and carbon monoxide levels. Similarly, a study from Taipei concluded that the likelihood of peptic ulcer hospitalisations rose significantly with increases in air pollutants during the study period.

Gut cancers also may be associated with increased air pollution with some evidence for stomach cancer and colon cancer as well as the liver cancer described in the previous section.

Climate Change and obesity

The rising tide of obesity worldwide is becoming a major problem and experts predict a relentless increase in this problem. Climate change impacts upon obesity levels via a number of mechanisms. Not only that, but obesity presents yet another feedback loop that makes climate change worse.

Obesity is now described as being at epidemic levels. It affects more than 680 million adults worldwide and is present in nearly 40% of the US population. Globally, obesity has nearly tripled since 1975 and most of the world's population live in countries where being overweight or obese kills more people than being underweight. Obesity is associated with a risk of various malignancies, including breast and endometrial cancers, cancer of the oesophagus, gastric cardia (stomach), colon, rectum, liver, gallbladder, pancreas, kidney, thyroid gland and multiple myeloma. It is a major risk factor for Type 2 diabetes, heart disease, lung disease and early death. It significantly reduces quality of life, social interactions and mental health. I could go on.

As the planetary temperature graphs have risen in a similar way to the obesity graphs, researchers have been looking for a possible link. I have summarised the findings below but for detailed information on the research have a look at The Lancet Commission on Obesity report in 2019 entitled 'The global syndemic*[1] of Obesity, Undernutrition and Climate Change'.

---

[1]   A syndemic is a relatively new word (coined in the 1990s) that is used to describe two or more diseases that act together to make each other worse. The definition can be broadened to include environmental factors as well as biological disease.

The effect of climate change on obesity: Increased temperatures and humidity inevitably mean that people go outside less, preferring to stay indoors in shade and air conditioning. This means people are less physically active, and so expend fewer calories.

As weather events get more extreme, fruit and vegetable crops fail and these products become more expensive. Processed food and drink which generally use cheap sugar, salt and oils, become relatively much less expensive. Processed food is a major risk factor for obesity.

Economic issues related to climate damage increase food insecurity. Those who are financially struggling use cheaper, calorie rich but nutritionally poor meals to reduce outgoings. This effect has been noted to have a particularly adverse effect upon children and is found in rich and poor nations.

Increase in environmental heat, weight gain and reduced exercise tolerance make car transport far more likely. This further reduces physical activity. Increased levels of air pollution have also, perhaps unsurprisingly, been shown to reduce exercise amounts and reduce physical activity. Smog is not going to tempt anyone out of their front door.

Poor sleep – an often overlooked feature of climate change – is caused by high night-time temperatures and humidity (wet-bulb temperature – see Chapter 1). Insomnia is responsible for a number of ills but overeating and weight gain have been convincingly linked to poor sleep quality and quantity.

The effect of obesity on climate change: Those who are obese have around a 20% heavier climate footprint than those of average weight, of which 52% comes from food and drink consumption. Diets of obese individuals usually have 30% more calories as they require more energy to maintain their greater body weight. In 2015, excess body weight was estimated to affect 2 billion people worldwide and it is calculated that every dietary calorie (1 kcal or 4.184 kJ) is equivalent to 2.21g of greenhouse gas emissions. The **total impact of obesity** on climate change may be an extra 700 megatons per year of CO_2 emissions.

The agricultural system is driving unprecedented environmental damage, accounting for 29% of anthropogenic greenhouse gas emissions and causing rapid deforestation, soil degradation and massive biodiversity loss. Meat production is at the centre of these costs. One study in 2016 demonstrated that high meat availability is the best predictor of obesity prevalence.

Ultra-processed food is an environmental, as well as an individual's, nightmare. It is responsible for 17-39% of all diet-related energy use, 36-45% of total diet-related biodiversity loss, up to a third of total diet-related greenhouse gas emissions, land use and food waste, and up to a quarter of total diet-related water use in its preparation. The reduction in physical activity (walking mainly) in the obese results in an increase in car usage with further greenhouse gas emissions from these vehicles. The increased use of home air conditioning in sedentary individuals plays its own role in adding to emissions.

If you are overweight or obese you can see that losing weight is a win for you and your planet.

Climate Change, dementia, and other neurological diseases

Our neurological system consists of our brain, our spinal cord and the myriad of nerves that cover our whole body that allow us to sense and react to our environment. Diseases that affect any of these structures usually have a profound effect on the sufferer's life. As human populations age, neurological diseases increase, and we are all aware (and constantly reminded by News items) of the immense impact of dementia and strokes. There are also a whole host of perhaps lesser known, but equally debilitating, neurological conditions that you may or may not have heard of. That climate change could increase the risk of these life-changing diseases is a frightening thought. But does it?

The most comprehensive answer to this question comes in a large systematic review in the journal *Neurology* in 2023 that looked at the impacts of climate change and air pollution on neurological health. They found 364 studies published in this area. The detailed findings are below.

<u>Dementia (Alzheimer's Disease).</u> The strongest association of climate change and dementia seems to be, once again, via air pollution. Over 6% of dementia cases are estimated to be directly attributable to particulate and nitrous dioxide exposure. Clinical diagnosis of dementia has been associated with years of exposure to airborne pollutants. Brain scans show greater areas of loss in sufferers who have been exposed to more years of these pollutants.

There is research evidence that heat may also have an adverse effect upon proteins in the brain – disordered brain proteins are a part of Alzheimer's Disease. Similarly, heat has been shown to cause an excess of certain so called excitatory amino acids in the brain. These excitatory amino acids have been implicated and backed up experimentally to be linked to future dementia. This research is at an early stage and remains speculative – though would provide a mechanism for the possible heat and pollution damage and Alzheimer's.

Geographical studies have shown a link between living near a major road and the risks of having a dementia diagnosis. A Canadian study found that those who lived within 50 metres of a busy road had the highest increased risk, although those who lived up to 200 metres away also saw a slightly increased risk. For those living 50m away, the risk was 7% higher than those living 300m or more away from the road. For those living within 200m, the risk of developing dementia was increased by 2%.

To add to this burden, hospital admissions for patients with dementia increase at higher temperatures. One study showed mean temperature increases of 1.5°C increased hospital admissions by 12%. Wildfires seem to increase this risk even further with a marked spike in hospital admissions during and shortly after the fires.

Stroke. A stroke is a sudden decrease in blood supply to all or part of the brain usually caused by damage to the vessels carrying this blood. Depending upon the amount of brain damaged and length of time the damage occurs, outcomes can vary from limited functional loss to severe loss to death.

As we have already seen, diseases of the blood vessels are more likely when the person is exposed to high levels of airborne pollution. One analysis concluded that 9% of stroke disability and 8.5% of stroke deaths could be attributed to particulate exposure. Another study found a nearly 50% increase in the risk of stroke for those living within 100 metres of a major roadway. A Chinese study found a link between very small particulate exposure and increased risk of stroke for three years after exposure. A German study found the same but the risk was for all polluting particle sizes.

There is also a direct increase in the risk of stroke with increasing environmental temperature and humidity – thus, high wet-bulb temperatures present the highest stroke risk. A Chinese study attributed 2-4% of all strokes to temperatures 5-8°C higher than usual for that location and time of year. An astonishing South Korean paper indicated that the major disability burden from heatwaves was from stroke.

Headaches. These vary from the barely noticeable to the quality of life destroying. There are a huge number of causes of headache, but the vast majority have no underlying serious cause. Headaches that result in medical attention have been shown to be more common with raised temperatures in the preceding days. Migraines, which are a particular type of headache (likely related to blood vessel over-sensitivity), have also been shown to increase during warm weather and high humidity. There is also some evidence that headaches – migrainous and non-migrainous – are increased when high levels of environmental air pollution are present.

Epilepsy. The associations are less clear for seizures. Some studies have indicated that unstable weather can increase the risk – with greater fluctuations in temperature and humidity appearing to increase the number of fits. Raised ambient temperature itself has not been shown to be a specific risk. Severe heatstroke can lead to fitting, but this is a different mechanism from the common epileptic seizures.

Multiple Sclerosis (MS). This a variable disease that in some sufferers causes major disability. It characteristically progresses through exacerbations over months and years. Higher temperatures than normal for that region have been recognised for some time to increase severity of MS. This is confirmed in various studies with one showing an 8.8% increase per 1°C increase in temperature.

Short term exposure to high air pollution also seems to have an effect on exacerbations but no link has been found, as yet, with MS development.

It is also important to note that MS, as with a number of neurological conditions (and also diabetes), can affect the sufferer's ability to regulate their body temperature. Our nerves detect and allow a response to raised environmental temperatures so when there is a dysfunction this can reduce the body's response to heat. Thus, those with neurological problems might be at greater risk of serious harm during heatwaves.

Parkinson's Disease. This is a relatively common disorder of movement with the sufferer having great difficulty in initiating movements as well as tremor and body rigidity. A study has shown an increased risk of Parkinson's in those exposed to Nitrous Oxide, Ozone and particulate pollution over a two year period prior to diagnosis.

Motor Neurone Disease (also called Amyotrophic Lateral Sclerosis or ALS). This is a degenerative disease of the muscles, leaving the sufferer unable to move or eat and eventually needing to have lung support. A study has shown an increased risk of MND/ALS in patients with long-term exposure to particulate pollution – though this has not been replicated as yet. It is thankfully a rare disease but there has been much research on its possible causes – a link with increased air pollution would certainly seem a worthwhile avenue for future research.

Brain Infections. Yet again, there is good evidence that the microorganisms that cause disease in humans are already thriving under raised environmental temperatures as well as the other opportunities climate change brings them. As with other infections, brain infections are becoming more common and extending their geographical spread. Some of the commonest examples of organisms that cause brain diseases such as meningitis and encephalitis are West Nile virus, Japanese encephalitis virus, Tick-borne encephalitis and coccidioidomycosis have been shown to be becoming more common. We will discuss these in more detail later.

Worm and parasite brain infections have not been shown to have increased, but it seems highly likely they will do so in the near future as conditions such as interruptions to water supply and soil run-off increasingly favour their spread.

Heatwaves and the Brain. We have seen already the damage that high heat and humidity can do to humans and the effect it has on the brain. It can cause strokes and it can disrupt the brain function in more subtle ways. It can change our ability to perform tasks, it can make us more irritable and aggressive. The increased heat causes stress hormones like cortisol to be released by the body and these have a direct effect upon the brain increasing our anxiety and stress. There is also some evidence that heatwaves make our brain even more sensitive to vehicle pollution as the barrier between the brain and the circulation becomes more porous.

It is unsurprising that dementia sufferers show a raised risk of health exacerbations when external temperatures are high. One study showed that a 4.5% increase in risk of dementia medical admission was observed for every 1°C increase in temperature above 17°C. They also calculated that, with a continued high emissions scenario, heat-related admissions would increase by almost 300% by 2040 compared to baseline levels.

As well as the direct effect of the heat, having to move dementia patients from their normal, familiar circumstances increases their risk of medical problems. Damage to care facilities from wildfires may mean en masse removal of residents – a risk factor for worsening health in this vulnerable group even without the other factors.

Climate Change, bone health and fractures

Our bones are not static structures like steel girders in a building, but complex, living tissues. Bones are constantly being remodelled – bits being taken away and bits added – depending upon our health, nutrition and the stresses we put upon them. They respond to personal and environmental changes just as other parts of the body do and are therefore just as much at risk from climate change as the rest of our working parts.

If our bones are weakened, they will obviously break more easily – we call this a fracture. Fractures can be big and obvious and very painful, for example a leg bone after a collision at football. Fractures can also be small and painless – these are called microfractures – and they can accumulate in one or more bones which can then eventually crumble like chalk. This type of fracture is more common as we age and especially in women after the menopause. Bones naturally become less strong over time (accelerated again by the female menopause) and this weakening of the bones is called osteoporosis. Osteoporosis is a major risk factor for fractures in older people. Fractures of any sort can result in pain, loss of function and interfere significantly with daily living. Anything that increases the risk of osteoporosis – in women or men – increases the risk of death or disability even with minor falls.

In the USA alone, around 2 million osteoporosis-related bone fractures are reported each year, resulting in as much as US$20 billion in annual direct health costs. Within one year of a bone fracture, death risks for older individuals increased by 10–20% with only 40% regaining full pre-fracture independence. Osteoporosis impacts women more than men, with 80 percent of the estimated 10 million Americans with osteoporosis being women. Postmenopausal women are at highest risk, with one in two women over 50 experiencing a bone fracture because of osteoporosis.

Climate change affects our bones in two different ways – reduced outdoor activity and air pollution. As we have already seen, raised external temperatures reduce outdoor activity such as running and walking. It creates more dependency on car use – even for short journeys. This has two deleterious effects on bones. Firstly, bones get stronger when mechanical stress, such as walking or running, is placed on them and conversely become weaker (this means they lose calcium) when these stressors are absent i.e. when you sit on a chair all day. Secondly, bones need an adequate supply of Vitamin D if they are to be healthy and strong. Vitamin D is made in the skin when the sun shines on it – sitting looking out of a window is not an adequate replacement for this.

The second climate related factor that affects our bones is that invisible villain again – air pollution. A study by Prada *et al.* found that the risk of bone fracture hospital admissions at osteoporosis-related body sites was greater in areas with higher particulate pollution levels. This risk was particularly high among low-income communities. Black carbon (i.e. large particle pollution) concentration was associated with higher bone mineral density loss over time at multiple anatomical sites than would naturally occur.

Another study found that elevated levels of air pollutants were significantly associated with bone damage among postmenopausal women. The effects were most evident on the lumbar spine, with raised nitrous oxide levels twice as damaging to this area than would normally be seen with ageing. The researchers analysed data collected through the Women's Health Initiative Study, an ethnically diverse cohort of 161,808 postmenopausal women. They estimated air pollution exposures based upon participants' home addresses. They measured bone mineral density at enrolment then follow-up at year one, year three and finally at year six. The magnitude of the effects of nitrogen oxides on lumbar spine bone density would amount to a 1.22% reduction per year which is nearly double the annual effects of age. This was shown for any of the anatomical sites evaluated in the study.

Sometimes climate change may have a positive effect upon diseases and disability. Warming of the planet is likely to reduce certain types of fracture. Going outside less and exercising less will reduce the risk of outdoor falls and therefore fractures. As ice and snow days become fewer there is consequently less chance of slipping. These factors are nothing to do with bone health, just some of the precipitating factors that might act on the weakened bones. By sitting watching TV, we are not outside falling over but we are day by day weakening our bones. Which would you choose?

Climate Change and dental health

Climate change can have direct effects upon our oral health, our teeth and dental disease. Dental caries are more common with higher temperatures and humidity due to the promotion of the mouth bacteria that cause this damage to teeth. Reduction and change to water supply can further exacerbate this effect.

Oral cancers have been linked to climate change induced increases in UV radiation (so no different to our outer skin). Gum infections (periodontal disease) are also more likely in hotter and more humid conditions – again because of the promotion of microbacterial overgrowth. Changes in nutrition precipitated by climate change can have profound effects on teeth, gum and general mouth heath. Lack of vitamins and minerals from stressed plants causes the same problems in our mouths as it does on the rest of our body. So yes, climate change can even cause bad breath!

As with other health professionals, dentists will be impacted by climate change with a greater patient burden caused by the diseases above. As with other professions, dentists have a duty to look at the effect they have on the climate and sustainability. From encouragement of good dental health to disposal of waste to reducing energy consumption, all are needed to reduce the impact upon the planet.

Climate Change, fertility and pregnancy

Many aspects of reproduction seem to be adversely affected by climate change and this seems to be particularly related to heat stress. Several studies have shown that environmental heat can reduce fertility. One study showed a nationwide decrease in births of 0.4% nine months after a heatwave. An Italian study indicated that a 1°C increase in maximum temperature decreased the overall fertility rate. Animal studies have suggested the reason for this is related to increased temperature causing menstrual cycle changes and alterations in reproductive hormones.

Sperm production is closely related to temperature, which is why it is manufactured external to body cavities – we should not be surprised therefore if higher environmental temperatures have an effect. A meta-analysis of the impact of raised ambient temperature on human sperm showed that high environmental temperature negatively affected sperm quality, including decreased semen volume, sperm count, sperm concentration, motility and normal morphology.

It is well known that worldwide fertility rates are declining and there are likely to be many reasons for this (including improved female education and literacy which reduce the number of children women have; male literacy makes no difference) but one reason not often mentioned is rising global temperatures. Perhaps it is nature providing feedback – preventing us sending more and more humans into an uncertain future. Whether or not that is the reason we have already seen that, with the declining fertility in our warmer world, the future is already here.

There are some other reasons that fertility might be reduced with more extreme climate events. Behavioural changes are likely – poor harvests have already been shown to increase contraceptive use. Crop failures, the shock of flooding, the stress of drought or the poison from wildfires are making those affected think of both the economics of more children and perhaps, the wisdom of bringing children into such a frightening world.

Once pregnancy has been established, the effect of climate change on the mother and unborn child bring their own threats. Pregnant women are more prone to heat stress than non-pregnant women due to their compromised thermoregulation and homeostasis ability. As we might therefore expect, there is evidence that there is a relationship between climate change related exposures and adverse pregnancy outcomes. These include eclampsia and preeclampsia, preterm birth, low birth weight babies, pregnancy induced raised blood pressure, and increased maternal and baby mortality. It has also been estimated that 25,000 infants per year between 1969 and 1988 were born earlier than normal as a result of heat exposure.

The WHO estimates that 88% of the burden of diseases attributable to climate change occurs in children under five years. A 2020 meta-analysis of 70 studies across 27 countries examined the impact of high temperature on preterm birth, low birthweight, and stillbirth. They found a 16% higher risk of preterm birth during heatwave days compared to non-heatwave days, with each additional degree Fahrenheit (0.56°C) associated with a 5% increased risk. Additionally, they found that the low birthweight rate was 9% higher during periods with hotter than usual temperature with babies born, on average, 26 grams lighter. Stillbirth risk was 46% higher during heatwave compared to non-heatwave days, with a risk increment of 5% for each additional degree Fahrenheit.

There is a substantial amount of evidence that air pollution harms expectant mothers, the foetus and the new-born. A South African study found that exposure to particulate pollutants had both significant direct and indirect effects on the risk of **all** adverse birth outcomes. Similarly, an increased level of maternal exposure to atmospheric Sulphur Dioxide during pregnancy was associated with an increased probability of preterm birth and the baby being small for gestational age.

Air pollution may permanently affect lung development in the baby through low birth weight, early birth, or incomplete immune system development. The health implications of this exposure are especially important as air pollution during the prenatal period may interfere with organ development. Infant and child lung function impairment, an increase in respiratory symptoms and the advent of childhood asthma have similarly all been linked to prenatal exposure to air pollution.

Pregnancy and early childhood are critical times for the formation and maturation of body systems, and the time during which the most rapid changes take place. Factors that adversely affect human development, including air pollution, can have both immediate and long-lasting effects on a person's health and some of these health impacts may only emerge later in life. We can't fix climate change in the short term but we can fix air pollution. We now know that this pollution is affecting the very youngest children and these effects might last a lifetime. We are at the beginning of this research and, by definition, it might take years to know the full implications. Do we really want to wait to find out?

Climate Change and eye diseases and blindness

As with our skin, our eyes interface directly with the external environment. They also, like the skin, already have a range of diseases that are caused by excessive sun rays, heat, dryness and wind. The eyes are therefore one of the organs that will be the first to suffer from our changing climate. As we will see, there is already evidence for this effect and the potentially disastrous consequences of eye disease and blindness in an already more unstable world. This increase in eye diseases will be found everywhere – not just in tropical latitudes, and we all need to be worried about the problems this will bring.

Drier atmospheres contribute to disorders of the front, transparent 'window' of the eye – the cornea. The cornea is clear to allow light to pass through it and onto the retina so any damage to the cornea can have a very significant effect upon vision. Solar radiation, high temperatures and the drier conditions have long been known to cause damage and scarring of the cornea – a condition called a pterygium. These are far more common in warm, arid countries and outdoor workers such as farmers have a far higher risk of getting them. Most can be surgically removed if that surgery is available in that country. Many recur and can leave a previously transparent tissue with a permanent opacity. The warming, drier, irradiated world we are facing will see an epidemic of pterygia and thus a significant increase in corneal blindness.

Sitting behind the cornea and essential to focus the light rays correctly onto the back of the eye is the lens. The natural lens is similar in function to the lens of a camera and like a camera lens needs to be crystal clear to allow the best images to be formed. Light and especially UV light is energy and the more energy going through a tissue the more damage that will occur. As we saw previously with skin disease, the amount of UV radiation hitting the Earth is increasing as the Ozone layer thins, and this is compounded by higher atmospheric temperatures. This increased energy within the lens disrupts its proteins so making it cloudy – this is what we call a cataract. Although cataracts are amenable to removal and replacement with a prosthetic lens, this is not always readily available in some countries. In Africa and parts of South Asia, where cataract prevalence is highest, surgery may be too expensive or require too much travel or there may simply not be enough eye surgeons to serve the whole population.

This is unfortunately true for other eye diseases and this can lead to reversible and irreversible blindness. As well as the personal burden these sufferers bear, there is a significant economic one. Poor sight represents a severe financial burden if you are unable to work, tend your crops or fetch water. This is not only for the individual, but for their families as they then must look after an economically inactive person. As our climate changes in the ways we have seen, we leave behind those without enough income to escape the changes. For those with visual impairment, the results of climate change are even more burdensome.

The increase in infectious and vector-borne diseases are, once again, going to increase in numbers and geographical spread as the Earth warms. In tropical and sub-tropical areas, infection related eye diseases and blindness already take a significant toll. There is a long list of infective agents that cause eye diseases but the two commonest are Trachoma and Onchocerciasis. Trachoma is an infectious disease caused by *Chlamydia Trachomatis* (this is different from the sexually transmitted chlamydia) and is spread by poor hand hygiene (worsened of course when water is scarce). It causes inflammation of the eyelids and their lining which is initially very irritating for the sufferer. As it progresses, it causes more and more inflammation which leads to scarring. Scarring causes the eyelids to turn in meaning the eyelashes abrade the cornea – which is extremely painful. If this is not treated, permanent damage to the cornea will occur and therefore visual impairment and even blindness.

Onchocerciasis or *River Blindness* occurs in tropical and sub-tropical regions and is caused by a parasitic worm that gains entry to humans via a black fly vector. These flies commonly bite humans and this is where the worm enters the body. It then causes skin problems, but when it reaches the eye via the bloodstream it can cause severe inflammation and damage to the optic nerve. Once again, severe cases will lead to blindness unless treated. Changes to water supply and warmer conditions encourage the flies that cause river blindness, whilst the worm can withstand harsher environments.

Air pollution is, yet again, linked to a variety of diseases, and the eye and visual system are no different. Poor air quality has been linked to damage of the cornea and surface of the eye, increased cataracts, higher incidence of glaucoma and of dry eye. The blood supply to the eye is substantial and, as we have seen, the blood vessels and vascular system are adversely affected by air pollution. Blood clots can travel to the eye in the circulation, blockages of blood vessels can occur inside the eye itself and the large parts of the brain that deal with vision and images can be affected by stroke. Any of these conditions can – and do – have a catastrophic effect on the vision and therefore of the sufferer's daily functioning.

One of the common issues that climate change has already brought and will accelerate in the forthcoming years is crop failure and subsequent nutritional deficiencies. As we have already discussed, the loss of plants from warmer temperatures, soil erosion, floods and droughts will more than outweigh the increased plant growth from raised atmospheric CO_2. Vitamin and mineral deficiency from this poorer diet can cause serious eye problems. Vitamin A deficiency is probably the most common and can lead to blindness (called xerophthalmia) especially in young children. Deficiencies of vitamin B1 and B12, vitamin C, vitamin D, and vitamin E, and minerals such as zinc can cause a wide range of eye problems from the mild to the sight threatening.

The thin tissue at the back of our eyes that allows us to see – the retina, is also susceptible to our changing climate. Age Related Macular Degeneration (ARMD) is a problem of the very centre of our vision and so affects reading and fine work. It is one of the commonest eye diseases all over the world and, due to our ageing population, the prevalence is increasing (as are the numbers who are being registered blind from it). Although increasing age is the biggest risk, there is evidence that other factors can play a part. Vitamin deficiency, especially of Vitamin E, has been shown to promote ARMD, similarly it is thought that zinc deficiency increases the risk. Increased UV light is also likely to be a promoter of this condition.

A different, but equally serious, retinal condition occurs when it peels away from the back of the eye – this is called a retinal detachment. Retinal detachments need surgery to replace the retina and this needs to be done as soon as practical. The surgery itself is one of the more complex eye operations and needs expensive, specialist equipment. In some countries, the operation is only available in the larger ophthalmology centres in the cities. Retinal detachment has been shown to occur significantly more frequently when temperatures were higher in the preceding week.

We discussed in an earlier section the increased risk of cancers as climate change progresses. The eye is no different in this regard. The commonest tumour arising from the eye is a melanoma. These tumours can not only create a blind, painful eye, but they commonly spread locally and in the blood stream as metastases. These ocular melanomas are increasing, and this is thought to be a combination of increased UV radiation and the higher temperatures.

The human cost of blindness is immeasurable. The economic cost to that individual and their family might be a matter of life and death. The economic burden to the region or country of an increase in those who are visually impaired or blind is, and will, be substantial. Worldwide it has been calculated that there are over 160 million people with moderate to severe visual impairment or blindness who are within working age with an estimated overall relative reduction in employment of people with vision loss of over 30%. The annual cost of potential productivity losses of visual impairment and blindness is $410.7

billion which is 0.3% of GDP. Overall productivity losses were estimated at over $408.5 billion.

In the USA, the total US economic burden of vision loss and blindness is over $134 billion. This comprises nearly $100 billion in direct costs – medical, nursing home, and supportive services. Over $35 billion in indirect costs – work absenteeism, lost household production, reduced labour force participation and informal care. The largest components of costs are medical costs ($53.5 billion), nursing home costs ($41.8 billion), and reduced labour force participation ($16.2 billion). Nationally, vision loss and blindness cost an average of $16,838 annually per person affected. A study in the UK found that the value of the loss of healthy life associated with sight loss and blindness estimated it to be over £19 billion.

The costs of eye disease, its treatment and the consequences when that treatment fails or is not available are immense. We can see how much two rich Western countries already spend on eye conditions, and we know this figure will inexorably rise. In developing nations, these costs are much less even though they have more preventable sight loss simply because there are not enough facilities or income to spend on treatment. This only means that the personal, the human cost, is so much greater.

Climate Change and insomnia and its consequences

The threats to our health from climate change may be subtle and easily overlooked. Chronic insomnia is one of these threats. Chronic insomnia is defined as the inability to have refreshing sleep at least three times a week for at least three months and which cannot be fully explained by another health disorder. In 2015, the US Centre for Disease Control and Prevention declared sleep disorders, and insufficient sleep in particular, to be a public health epidemic. On average, most people need 7-8 hours of sleep to feel fully rested. The National Sleep Foundation found that in 2017, only 47% of working Americans achieved this. This was even lower in Japan with only 34% sleeping more than seven hours. It has been estimated that in the USA, 50-70 million people are affected by sleep disorders. There is also some evidence that sleep problems are becoming a problem in Africa and Asia with an estimated 150 million adults in these areas suffering.

Awareness that insomnia is more than an inconvenience, but a real threat to health is increasingly recognised. Studies have found an association between chronic insomnia and diabetes, raised blood pressure, heart attacks, strokes and dementia. Chronically poor sleep is strongly associated with depression, anxiety and other mental health issues. It is an independent risk factor for suicide. The rate of alcohol dependence is much greater in those who report poor sleep. Accidents are more common in chronic insomniacs, especially driving incidents. Overall, it has been shown that those who get fewer than five hours sleep per night have a 15% higher mortality rate than those who sleep for more than five hours.

One of the greatest threats to health from sleep disorders is sleep apnoea. This is a disorder in which, when asleep, the breathing passages close and prevent air getting to the lungs. This reduction in oxygen wakens the person from sleep and they then catch up a breath by opening their airway (they produce a loud 'grunt' when they do this). It is likely that far more people have this condition than are diagnosed, but it has the highest association with death and disease of all the sleep disorders as well as the greatest risk of daytime drowsiness and accidents. Sleep apnoea is closely linked with obesity.

Sleep disorders also have an economic impact upon the individual and the society. Poor work performance, mental health issues and ongoing medical needs increase the risk of unemployment. Chronic ill health is expensive in other ways for the individual with more medications needed and more medical care. In the USA alone more than 100,000 car crashes, 1550 deaths, 71,000 injuries and over $12 billion monetary losses are attributable just to driving accidents when drowsy. If this is multiplied by other health-related outcomes, then worldwide the economic burden to society of poor sleep is enormous.

A significant number of studies have shown a link between climate change and insomnia. These effects are from a number of overlapping factors. Higher temperatures and humidity – especially at night – have been found, in numerous studies, to reduce sleep time and quality, and to increase the risk of sleep apnoea. A $1°C$ increase is associated with a greater number of people having more than three nights of insomnia. To add to this, it has been shown that sleep disruption itself can also disrupt the body's ability to regulate its temperature.

We have seen earlier the profound effect of obesity on ill-health. Obesity and insomnia are closely linked. We have also seen that climate change makes people more likely to be obese – from poorer diet from agricultural failures to decreased activity in the hot, polluted air. Obesity itself worsens climate change, mainly through increased use of energy for cooling and transport.

Studies have shown that there is an increase in sleep-related disorders following extreme weather events. The main work has been done in relation to hurricanes but there is also evidence that this effect is present after flood and wildfire events.

Air pollution once again rears its ugly head. A study from China looked at the association between ambient air pollution exposure and insomnia among adults in Taipei. This showed that average particulate pollution, raised ozone levels and raised Nitrous Oxide levels were associated with a higher number of insomniacs. After adjusting for confounding factors, an increase of only one unit in the one-year average level of particulate matter showed a statistically significant association with insomnia. Warmer, humid, polluted areas reduced residents' sleep quality and length, and increased their risk of health issues. These climate conditions are of course exactly those we expect with increased global warming.

Our increasingly 24-hour societies and concurrent use of electronic devices can profoundly disrupt our sleep patterns. We are more sedentary as we are more sleepy during the day so use our electronic devices to reduce energy expenditure movements further e.g. food delivery, face to face social interactions. We use air conditioning – in our houses or vehicles – to simulate the coolness that we used to get by going outside. All these things need external energy to work and therefore increase the use of fossil fuels to run them. Yet again we create a climate problem for which the 'solution' worsens the climate problem.

Climate Change and vulnerable populations

We have already seen that although climate change will affect all of us, certain populations are more vulnerable to its effects.

Geographically those in warmer climates are going to suffer from the earliest climate changes and indeed already are with increased rates of crop failure, infectious diseases, floods and droughts in countries nearest to the equator. As has become obvious, these problems are slowly but inexorably moving into previously cooler climes.

Even in these high risk areas some people are far more vulnerable than others – in particular children, the very elderly and the poor. The young and the elderly both have poorer temperature control systems than healthy adults, making them much more vulnerable to even small temperatures changes. Infants have a high surface area which means they absorb heat more quickly when the air temperature rises. The elderly have a diminished sweating ability which, as we saw in Chapter 3, reduces their ability to cool down. Both groups have less reserve in their heart and lungs, thus particulate matter from vehicles or wildfires can disproportionally damage them. They also need much tighter control of their fluid balance so are again disproportionately affected by drought situations.

Both groups have fewer behavioural levers to control their temperatures, usually relying on others either to regulate the room temperature or to move them away from warmer environments – temporarily or permanently. Children rely upon parents and carers to protect them from high heat – sadly, parents are not always able to do this. The elderly are often more socially isolated and this is a big risk for heat-related deaths. The economics of ageing frequently mean the elderly have less access to air conditioning or are more reluctant to use it.

Figures from The United Nations at the United Nations Framework Convention on Climate Change (UNFCCC) are startling:

> 240 million children are highly exposed to coastal flooding.
> 330 million children are highly exposed to river flooding.
> 400 million children are highly exposed to cyclones.
> 600 million children are highly exposed to vector-borne diseases.
> 815 million children are highly exposed to lead pollution.
> 820 million children are highly exposed to heatwaves.
> 920 million children are highly exposed to water scarcity.
> 1 billion children are highly exposed to exceedingly high levels of air pollution.

Remember that these are the figures at present and in the future will only get worse. As Henrietta Fore, UNICEF Executive Director says:

> For the first time, we have a complete picture of where and how children are vulnerable to climate change, and that picture is almost unimaginably dire. Climate and environmental shocks are undermining the complete spectrum of children's rights, from access to clean air, food and safe water; to education, housing, freedom from exploitation, and even their right to survive. Virtually no child's life will be unaffected.

We are only at the beginning of our understanding of how climate change is going to affect our and our children's health. For example, an analysis of nearly 16,000 child and adolescent or young adult patients with cancer in Utah revealed that exposure to fine particulate matter was associated with increased mortality at 5 and 10 years after diagnosis of certain cancers.

The elderly will face their own challenges in the warming world. As well as impaired temperature regulation they often have co-existing medical problems. Heart, lung, kidney and liver diseases all are worsened by climate change as we have seen, and all are much more likely to occur in the elderly. Medical conditions and often their medications (e.g. diuretics/water tablets which reduce the fluid in the body) can interfere with the body's temperature control and worsen any environmental stresses. The vast majority of city heatwave deaths are in the elderly – 70,000 people died in the Europe heatwave of 2003, by far the majority being over 75 years of age. This was similar in Russia in the 2010 heatwave when 55,000 died. The July 1995 heatwave in Chicago reported 514 direct heat-related deaths and 696 overall excess deaths with those 65 years of age or older overrepresented.

Older adults often have limited mobility, making it difficult to get to safety during extreme weather events. Extreme events can also interrupt their medical care, making it hard for them to be transported with their correct medications, medical records and health equipment. Power outages after a storm can also affect elevators, air conditioning or heat, and electronically powered medical equipment, making older adults particularly vulnerable. They suffer more from respiratory and heart disease meaning that wildfires carry a higher risk of mortality for them as well as mobility issues, reducing their ability to move from the smoke zones.

The poor of the world – be they young or old – are far more at the mercy of the climate than the rich. They can't use their credit cards to jump on a plane, and reach safety from heatwaves or wildfires or hurricanes or floods. No matter where you are in the world – developed or developing countries – if you are poor your options to deal with climate change are much more limited. Poorer people tend to live nearer to roads and industry (i.e. where other people don't want to live) so they are far more affected by air pollution. Failed agriculture increases healthy food prices much more than ultra-processed unhealthy food and so the poor consume cheaper, less nutritious food. The poor are also more likely to have to drink unhealthy water during droughts, increasing the risk of water-borne diseases. The list goes on as poverty, like climate change, affects all aspects of people's lives.

A perhaps overlooked group that are at greater risk of health issues from climate change are those who work outdoors. A farmer in Africa or a construction worker in Australia are both more vulnerable from climate change. Increased sun exposure, heatwaves, insect bites passing on disease, air pollution, weather events, water-borne diseases all present an increased risk to those who are not protected by buildings. Any local or national health strategy needs to bear this in mind when planning future services as the risks to these workers multiply.

Most of us are aware of the climate paradox (or at least one of them as there are a few) – that those who emit the least CO_2 and are least responsible for climate change are the ones who will suffer its worst effects. Research by Oxfam shows that the world's richest 10% of people are responsible for 50% of emissions. This group also claims over half of the world's wealth.

The world's poorest 50% of people contribute approximately 10% of global emissions and receive about 8% of global income.

Data from the World Bank shows that the average person in the UK emits 65 times more carbon compared to someone in Malawi. US, Canadian and Australian citizens emit over 150 times more. At the same time, the sixth poorest country in the world, Mozambique, shoulders the burden of over $3.2 billion in loss and damage following two unprecedented cyclones in 2019.

Climate Change and mental health

The majority of the health conditions affected by climate change we have discussed so far are purely physical. Mental health is also significantly affected by climate change. As our bodies have evolved to thrive in a particular climate so have our brains. As our familiar environmental parameters change – as we can't sleep because of the heat, as we see our neighbours migrating, as we are choked by wildfires or watch our crops fail again – these take their toll. The brain looks for familiarity, for the routine that signifies safety. When it can't find this, it puts us into stress mode as it assumes we are under threat. Anxiety, stress and their consequences are as potentially damaging to us as any physical harm.

Mental health issues take a number of often overlapping forms, most of which are worsened by the pressures of climate change. A study by Obradovich *et al.* coupled meteorological and climatic data with reported mental health difficulties drawn from nearly two million randomly sampled US residents between 2002 and 2012. They found that shifting from monthly temperatures between 25°C and 30°C to greater than 30°C increased the probability of mental health difficulties by 0.5%. 1°C of 5-year warming was associated with a 2% increase in the prevalence of mental health issues. Specifically, those exposed to Hurricane Katrina had a 4% point increase in adverse mental health scores.

A Chinese study found a similar association. They obtained daily hospital admission data for 'mental disorders', daily meteorological and ambient pollution data in Shanghai from January 2008 to December 2015. They found that a temperature above 24.6°C resulted in higher psychiatric hospital admissions. The higher the temperature the higher the number of admissions with around a 25% increase with temperatures above 33°C.

More specifically, studies have found an association between long term exposure to elevated temperatures and major depression. Belova *et al.* projected changes in suicide incidence across the United States in response to warming and found that 1–6°C of warming would result in up to 1,660 additional suicide cases annually, a 4.1% increase from baseline. A large meta-analysis of 114 studies found a 1°C increase in mean monthly temperatures was associated with an increase of suicide of 1.5%. They also found an increase of admission for mental illness of nearly 10% during heatwaves.

It is not just raised temperature that contributes to this increase in suicides – air pollution has once again also been linked. Braithwaite *et al.* found an association between long-term particulate exposure and depression, anxiety and increased suicide.

There is evidence of climate change increasing substance abuse – tobacco, alcohol and recreational drugs (marijuana, cocaine etc.) – and this seems to be for multiple reasons. Vergunst *et al.* have classified these reasons into five pathways:

- Psychosocial stress arising from the destabilisation of social, environmental, economic, and geopolitical support systems.
- Increased rates of mental disorders.
- Increased physical-health burden.
- Incremental harmful changes to established behaviour patterns. Alcohol abuse increases in hot weather both as an escape mechanism and because of use as an attempted thirst quencher.
- Worry about the dangers of unchecked climate change (called Solastalgia – from the Latin *solacium* (comfort) and the Greek *algia* – pain, suffering, grief). This appears to be an increasing problem – so called 'eco anxiety'. The perception of an existential threat has, unsurprisingly, created an increase in stress and anxiety disorders.

These pathways could operate independently, additively, interactively and cumulatively to increase substance-use vulnerability. The authors also note that young people face disproportionate risks because of their high vulnerability to mental-health problems and substance-use disorders and the greater number of life years ahead in which to be exposed to current and worsening climate change.

This latter point is echoed by Olson and Metz who state that the threats from climate change put children at risk of post-traumatic stress disorder, depression, anxiety, phobias, sleep disorders, attachment disorders and substance abuse. We should all remember when we sit in our cars, book our long-haul flight or turn the heating up instead of putting on a jumper, that the molecules of CO_2 produced will be with us and our children forever.

A final note on alcohol and another negative feedback loop for us to examine. Alcohol production is itself detrimental to the planet through its manufacture and transport, and to human health due to the development of liver disease and other alcohol use disorders. In Sweden, alcohol consumption is responsible for approximately 3% of dietary greenhouse gas emissions, or 11% in subpopulations with the highest reported alcohol intake. It is interesting to think that successful alcohol reduction strategies (e.g. minimum price legislation) would cut alcohol disease and death, and would directly improve the health of the planet.

Climate Change and exercise

Lack of exercise is closely correlated to the development of a whole range of diseases. It has been said that if exercise was a pill it would be the most widely prescribed medicine in the world, such are its health benefits. It is important to remember though that exercise doesn't just mean going to the gym or running a marathon, it is also walking to the shops or climbing the stairs instead of taking the lift. Constant small movements are just as effective (if not more so) as going to the gym in promoting good health.

This connection between climate change and exercise is illustrative of how the changing climate can affect us and how we in turn affect it. As will be clear if you have read things so far, climate change is affecting all facets of our life and all facets of our life are affecting climate change.

<u>How does climate change affect exercise?</u> As the temperature increases this produces a direct threat to those doing exercise. Muscle activity of any sort produces heat as a by-product and the more exercise we do the greater heat we produce. We need to dissipate this heat or we will rapidly disintegrate so we sweat profusely. We then lose fluid which we need to replace or we will become dehydrated. You will notice that these features are very similar to the external temperature rising, and so a combination of sustained exercise and ambient heat is potentially very dangerous.

There is no threshold of heat where it is safe to exercise or not exercise. It depends upon your health, heat tolerance, how high the external temperature is, how high the humidity is and how long you might be exposed to these. Even walking can be dangerous if the temperature and humidity are high enough. There are enough lurid stories in the news of hikers dying in heatwaves each season.

Air pollution brings an obvious risk to those who run or walk in urban areas. Exercise is meant to stress the heart and lungs and so strengthen them, but the additional stress of air pollution can trigger disease or even death. Studies have shown decreased athletic performance, increased number of asthma attacks, changes in blood parameters and exercise tolerance in polluted areas. One alarming study from China that looked at the effects of air pollution on exercise concluded that 'our findings strongly suggest that, in the presence of relatively high air pollution, the cardioprotective effect of physical activity is not only attenuated but is largely absent'.

Even swimming has a risk of heatstroke. We generate a great deal of heat when swimming but our sweat obviously can't evaporate. However, because the water is colder than our body temperature the excess heat is conducted away into the water. As the water warms this becomes less efficient. In some parts of the world water temperatures can reach 87° F (30.5°C) which is life threatening for a swimmer. Open water swimmers have an additional risk from infectious diseases. We have discussed previously the rise in water-borne infectious agents in a warmer world and organisms such as leptospirosis and *E.Coli* are going to present increasing risks.

Hikers and wild campers will be at risk of more extreme weather events from flash floods, wind events and heatwaves. Those who enjoy skiing or climbing will be at increased risk of avalanches as the heat makes the snow less stable. They will also be exposed to risk from the increased frequency and strengths of storms. Already, as the snow line goes ever higher up the mountains, skiers must travel to higher altitudes.

We have already seen how high external heat, humidity and air pollution significantly reduce people's overall activity. We can all recognise our lethargy on very hot days but if those hot days continue into hot weeks and months, reduced exercise levels begin to bring their own problems. Heat, obesity and activity are linked, and this link is going to strengthen as we warm. Raised external temperatures have been shown to reduce 24 hour movement behaviours i.e. we sit and we sit for longer. This is especially so for those already with chronic diseases and the elderly. There have also been reports of decreased physical activity after natural disasters, and this has also been linked to insomnia which itself is a risk for reduced 24 hour movements.

How does exercise affect climate change? We have already discussed that when heat encourages more sedentary behaviour, this results in greater reliance on fossil fuel derived energy sources. Use of private cars for transport is increased as is home air conditioning (as was noted in Chapter 4, air conditioning is responsible for the equivalent of 1,950 million tons of carbon dioxide released annually).

Long distance travel to take part in sporting events or skiing or hiking also take their toll on the planet. Tourism alone is responsible for 8% of the world's carbon emissions. Even traveling to watch other people play sport is harmful, with a calculation of global sport emitting 300-350 million tonnes of carbon per year – similar to the total output from Spain.

Even 'e-sports' can damage the planet! It has been calculated that the technology sector, from the powering of huge internet servers to charging personal devices, will consume as much as 20% of the world's electricity by 2030. Of course greater use of electronic devices results in lower physical activity...

Climate Change and the pharmaceutical industry

As our environment is changing we have already seen the impact this is having and will have upon our health. Diseases that are already common in western populations will be exacerbated and the infectious diseases that plague the developing world will grow. But as our planet warms these differences between north and south will narrow.

Inevitably this will mean a greater burden of worldwide disease and the pharmaceutical industry will need to respond to this. I guess if you work in the pharmaceutical industry, you will see this as an opportunity – however for the sake of our health, an ever expanding drug industry is perhaps not to be welcomed.

Demand will increase for a whole range of drugs. With more air pollution will come the need for respiratory medicines. With greater threats from bacteria, fungi and parasites will come the need for antibiotics – both the ones we have and the ones we need to develop (as we have already seen, antibiotic resistance will be increased with climate change). Vaccines will be even more necessary and novel ones will probably be needed as previously rare infections become more common.

Drugs that protect our heart and circulation, such as those used to lower cholesterol and blood pressure, will be increasingly required as heat and air pollution worsen. We know that cancers are going to become more common with climate change so supply and innovations in anti-tumour medications will be vital. Mental health issues are caused and exacerbated by climate change, and again supply and research of efficacious medicines will be vital.

As the climate and weather deteriorate, the pharmaceutical industry itself may have its own problems. Wind events or floods, drought or wildfires can damage manufacturing plants as well as the supply of raw materials. The whole supply chain from manufacture to the consumer could be threatened at any point by climate impacts. In 2017, the global medicines supply was interrupted when Hurricane Maria hit Puerto Rico – home to more than 500 medical product facilities and producer of 10% of all drugs consumed in America. The following year, Pfizer, Merck, and Novartis had to stop production at their southeast US operations as Hurricane Florence approached.

Pharmaceuticals and the pharmaceutical industry also have a directly negative effect upon our environment, and again this is going to worsen. In 2019, the pharmaceutical industry produced 48.55 tons of carbon dioxide equivalent for every $1m it generated. This places it above even the automotive industry in terms of contributing to global greenhouse gases. Further, the sector currently accounts for almost one quarter (23%) of global water usage.

Pharmaceuticals in the water supply – from manufacture to throwing away of medicines to metabolites excreted in urine – are measurable and can have serious effects. As our water supply dwindles and pharmaceutical use increases, the contamination of water supplies will worsen. We already know some effects this is causing in animals but do not as yet know the effects it might have on us.

The climate change bind is illustrated once again. The deleterious effects upon our health from climate change will make us ever more reliant on the pharmaceutical industry. This industry will itself be threatened by climate events as well as being a major contributor to our deteriorating climate.

Climate Change and infectious diseases

If you are squeamish you might want to look away now as some of these organisms and diseases are not pretty…

They are on the increase, and are slowly spreading into regions where they have never before existed as our warming planet creates conditions where they can thrive. Later in this chapter, we will see some of the diseases that those in the tropics and subtopics already have to deal with, and which those of us in cooler latitudes should begin to prepare for.

What is an infectious disease?

This might seem like an obvious question, but to understand why these represent such a threat we must first understand the organisms and how they cause disease.

We are surrounded by living creatures, the vast majority of which are too tiny for us to see without a microscope. Most of these live happily alongside us without causing any problems and many are important for our survival and well-being. A small number cause diseases when they come into contact with humans and these are called pathogenic organisms. These diseases range from mild, such as the cold virus, to life threatening such as the cholera bacteria.

It is important to remember that even usually harmless microorganisms can cause disease in the right circumstance. This is usually when the host's (that is us humans) defences are down, for example when malnourished or ill from another disease, or in the very young and old.

These infectious agents are very different from each other as you will see from the descriptions below. It is important to be clear about language – for instance, bacteria and viruses are often confused ('just a bug') but they are very different and climate change will affect them in different ways.

Viruses are the tiniest infectious agents, only seen with the very high powered microscopes. They are smaller than a single human cell and cause infection by entering that cell and hijacking it to produce more viruses before bursting out of the cell and killing it. Viruses can be attacked by our immune systems but are not killed by antibiotics as bacteria and fungi are. They cause a huge range of diseases in humans, other animals and plants. Covid is perhaps the best known virus now.

Bacteria, though microscopic, are far bigger than viruses. They are relatively simple organisms consisting of a single cell. They have evolved with humans and live all over our bodies and especially in our gut. Most are harmless or work with us for the benefit of both. As mentioned above, a small number directly cause disease in humans whilst others do not unless our defences are down.

Fungi are more complex organisms and exist all around us, for example mushrooms and stilton. We breathe some of them in, and are covered in them without realising. Most species do not cause diseases in humans but some do. Minor infections such as ringworm are fungal but more serious infections occur, once again, when our defences are down through illness or malnutrition.

Protozoa are single celled organisms with a more complicated structure and life cycle than bacteria or viruses. Some can live harmlessly in humans but a number cause diseases, such as malaria, toxoplasmosis and amoebiasis. We will discuss these in more detail later.

Worms (also known as helminths) or, to be more specific, parasitic worms, are much more complicated multi-cellular organisms. They live in soil and water, but when ingested can thrive in the human gut and in the bloodstream. They use the human host for protection and nutrition, and this situation can be long term but can also be life threatening depending on the worm and the health of the infected animal.

Vector-borne Diseases

The various infective organisms described above have various methods of getting into us. One common method is when they are transmitted by another creature which in this context is called a vector.

Vectors are living organisms that can transmit infectious pathogens between humans, or from animals to humans. Many are bloodsucking insects – when they bite an infected animal they also take in the microorganisms with the blood. When they then bite a different animal, this microorganism is injected directly into the animal's bloodstream. A very common example of this is malaria. Other animals can also be disease vectors, for example rats which can carry and transmit bubonic plague or bats that can carry rabies.

Just as climate change is encouraging infectious microorganisms, many of their vector hosts are also thriving and extending their range of habitats. Cold winters don't just mean you have the major inconvenience of scraping ice off your car, they often kill the insect vector's larvae in the soil, meaning reduced numbers in the spring. Warm winters mean more of these vector larvae survive. Hot, longer summers mean they can reproduce through more cycles and so produce more offspring. Deforestation and clearing of habitats move more of these vectors closer and closer to human communities.

Why does climate change increase the risk from infectious diseases?

Global warming is inadvertently creating ideal conditions for infective organisms to thrive. There are several reasons for this, all of which will worsen if our current trajectory continues.

Insects are cold-blooded so need heat directly from the environment to function. This means insects that carry disease are more common in the tropics and sub-tropics, hence vector-borne infections are far commoner in these regions. As the world warms these insects both multiply more successfully and extend the areas where they can thrive. Additionally, they usually have part of their life cycles in water so warmer, wetter conditions allow them to flourish even further. Research has found that warmer temperatures increase transmission of vector-borne disease up to an optimum temperature. Just as they carry different diseases, different mosquitoes are adapted to a range of temperatures. For example, carried by insect-borne vectors, malaria is most likely to spread at 25° Celsius (78° Fahrenheit) while the risk of zika is highest at 29° Celsius (84° Fahrenheit).

Many infective agents are passed from human to human – by direct touch, aerosols from coughing, sexual contact or indirectly via water in urine or faeces. The closer humans live and in larger numbers, the greater the chances of passing on infections. Poor water supply means water-borne pathogens are more likely to be concentrated in water, which may also have to be used for hygiene. Climate change forces migration of people and animals and often concentrates them in cooler areas; this concentration of people and animals again significantly increases the risk of passing on infections.

Malnutrition, which we have seen is an increasing consequence of climate change, weakens the host's defences making them more vulnerable to infections. Even organisms that would not cause serious infections can, with a malnourished weakened host, be life-threatening. Similarly, droughts not only increase the risk of infected water but reduce the individual's ability to fight any infection.

The various catastrophic events that climate change is bringing significantly increase the risk of serious infections. Any major disruption to a society will mean disruptions to water and food supply, reduced medical care and poorer hygiene. Survivors of the initial event must deal with these secondary consequences of the catastrophes including a huge rise in these infections.

The ease of modern travel carries its own risks. New viruses can travel round the world in hours (see COVID-19), insect vectors can survive long plane journeys and antibiotic resistant bacteria spread to new areas. Plane travel is not only responsible for a significant amount of the CO_2 emissions that are driving the warming, but for the rapid spreading of infective agents. Progress always comes at some sort of cost, though sometimes that cost may not be realised for many years.

The microorganisms themselves may thrive in these future conditions – warmer, wetter conditions are ideal for fungi and bacteria. Protozoa and worms also thrive in warmer conditions and adapt to drought often much better than their hosts can. Viruses, as they live in cells, are not affected directly by climate change but, as we have seen, are given more opportunities via increased vector activity and weakened host defences.

Our eternal battle with bacteria was significantly swung in our favour by the discovery of antibiotics from the 1940s onwards. As we saw in the previous chapter, increased temperatures, more humans in smaller areas and increased migration are going to accelerate antibiotic resistance. Research has shown that increased temperatures increase both the rate of bacterial growth and the rate of the spread of antibiotic-resistant genes between microorganisms. There is also evidence that increased particulate matter, particularly the very small particles, increases antibiotic resistance.

The battle we thought we had won is now heading back towards defeat. Data from 2013–2015 suggests that an increase of the daily minimum temperature by 10°C will lead to an increase in antibiotic resistance rates of *Escherichia coli*, *Klebsiella pneumoniae* and *Staphylococcus aureus* bacteria by 2–4% (up to 10% for certain antibiotics).

What infectious diseases are we potentially facing?

All the infectious diseases described in detail below are either already becoming more prevalent, or it is thought they are likely to become so under the influence of climate change. At present, where you live is a major determinant of which infections you are more likely to suffer but, as we warm, the areas these organisms can survive in will expand.

Mora *et al.* in a paper in *Nature Climate Change* looked at the human pathogenic diseases that could be worsened by climate change. They found that 58% (that is, 218 out of 375) of infectious diseases already confronting humanity worldwide have been at some point worsened by climatic hazards (16% were at times diminished). They described 1,006 unique pathways in which climatic changes, via different transmission types, could lead to pathogenic diseases.

Some of the most common and/or most serious infections from each pathogen group are described below. The list isn't exhaustive as there are too many diseases to describe (a frightening thought by itself) without being repetitive. Many have similar infection routes and climate change will increase their threat in generally similar ways. As we really don't know the outcomes of ever increasing climate disruption, it may be that one or more microorganisms not mentioned in this list will emerge as the greatest threat to us.

As I said at the beginning of this section, you might want to look away now. These horrors already exist and cause harm, and we should be hoping for a world where they are disappearing rather than thriving. But as one commentator states – they are coming for you.

<u>Malaria.</u> Malaria is a common tropical disease caused by a protozoan parasite *Plasmodium falciparum* (there are other Plasmodium species but this is the commonest infective one). It is transmitted when an infected female Anopheles Mosquito bites a human and it is injected directly into the bloodstream.

Early signs of infection are fever and flu-like illness, often intense shaking, headache, muscle aches, and tiredness. Nausea, vomiting, and diarrhoea may also occur. It can become chronic and then damages the red blood cells causing anaemia and jaundice which are manifest as tiredness, loss of appetite and chronic ill-health. Sometimes, if untreated, the infection can become severe and may cause kidney failure, seizures, coma, and death.

Globally, the World Health Organization estimates that in 2020, 241 million clinical cases of malaria occurred, and 627,000 people died of malaria, most of them children in Africa. Because malaria causes so much illness and death, the disease is a great drain on many national economies. Since many countries with malaria are already among the poorer nations, the disease maintains a vicious cycle of disease and poverty.

We have already seen that malaria is likely to be more common as the globe warms. The insect vectors will flourish in the heat, will increase their reproduction in more commonly occurring stagnant water, and more easily infect their human hosts in increasingly crowded conditions. It has been calculated that increased temperatures could lead to a 12-27% increase in malaria transmission, affecting 200 million individuals. A recent study reported that the mosquitoes that transmit malaria in Sub-Saharan Africa have moved to higher elevations by about 6.5 metres (roughly 21 feet) per year and away from the Equator by 4.7 kilometres (about three miles) per year over the past century. If you do the sums you can work out how near these mosquitoes have moved to you.

Lyme Disease – Lyme disease is the most commonly transmitted tick-borne infection in the United States and among the most frequently diagnosed tick-borne infections worldwide. In similar fashion to malaria, it is passed on via an insect bite. The underlying infective organism is a bacteria called *Borrelia burgdorferi* and this is passed to humans by bites from infected ticks. These ticks usually feed on the blood of deer so thrive where deer numbers are high, with both needing long grass to feed on and to protect them. Deer do not develop disease when infected with *Borrelia*.

Lyme disease is divided into three stages: early localised, early disseminated and late. The early localised disease is distinguished by the red ring-like expanding rash called Erythema

Migrans at the site of the tick bite. Sufferers also have flu-like symptoms, tiredness, headache, fever, and muscle and joint ache. About 20% of patients develop the early disseminated disease, with the most common symptoms being multiple skin lesions. Other symptoms of the disseminated stage are flu-like symptoms, enlarged lymph nodes, arthralgia (sore joints), myalgia (sore muscles), damage to the nerves around the face, eye inflammation and most seriously meningitis and heart problems. The most common manifestation of the late disease is a chronic disabling arthritis as well as prolonged, disabling fatigue.

The ticks that transmit Lyme disease have dramatically expanded their range in the northern United States and Europe. This is again because of the conditions which allow them to head further north – warmer summers and less severe winters. The warmer summers also allow the deer to live further north, allow the long grass to grow and shelter the ticks meaning more people are out walking in these areas.

Leptospirosis – not many people have heard of this nasty infection, although wild swimmers and rowers usually have. *Leptospirosis* or *Weil's Disease* is caused by the bacteria *Leptospira*. These bacteria live in small mammals – especially rats – and when the rats urinate in water the *Leptospira* are released. They also live in the soil and so heavy rain can wash them into rivers. They can survive in fresh water and infect the human after ingestion by drinking or accidental splashing but can also enter the body through open cuts.

The mild form of leptospirosis is characterised by fatigue, muscle ache, headache, fevers, headache, abdominal pain, light sensitivity, cough, nausea, vomiting and diarrhoea. In 5-10% of cases progression to more serious damage occurs. Liver and kidney damage can be so severe that these organs fail. Widespread inflammation can affect the lungs, the brain and the heart and the mortality rate for this group is very high.

Climate change means higher annual mean temperatures will enhance the growth and activity of *Leptospira* bacteria and at the same time lengthen the infectious season and expand the geographical distribution of the bacteria. Higher rainfall amounts and more moist conditions are linked to increased *Leptospira* bacterial growth and survival. Another important future climatic risk factor for leptospirosis infections is the increased frequency of extreme weather events. Heavy rainfall, storms and associated flooding events cause soil run-off that contains the bacteria so increasing water contamination. Poor soil health from previous droughts or deforestation further increases this soil run-off. This is all made worse by poor sanitation, inadequate healthcare and crowded conditions. Drought episodes also increase river swimming and bathing so making exposure to *Leptospira* more likely.

Dengue, Chikungunya and Zika – these three diseases are sometimes classified together as they are all caused by mosquito-borne viruses. This mosquito is sometimes called the Tiger Mosquito and is either *Aedes aegypti* or *Aedes albopictus*. Although transmitted by the same vector these diseases have different features and prognoses.

Dengue or dengue fever as it is also known, is caused by the dengue virus and transmitted by the bite from the mosquito. Dengue is now the most common mosquito-borne disease worldwide (at least 10 times the number of cases of malaria). It is an increasing risk in over 100 countries, mainly in Latin America, Asia and Africa with occasional (but increasingly frequent) outbreaks in Southern Europe and the Southern USA. In Brazil alone, there were more than half a million cases notified in 2021, including 240 deaths.

75% of dengue cases are mild and self-limiting but around 20% result in a high fever, rash, muscle and joint pains and stomach upsets. A small number, between 2 to 5%, have more severe symptoms which can be life threatening and without supportive treatment have a high mortality.

Chikungunya is caused by a virus of the same name. The commonest presenting symptoms of chikungunya are fever and joint pains, muscle pain, headache, nausea, fatigue and rash. The joint pain is very characteristic of this disease and can vary from mild to very debilitating and may last from a few days to many months or years. In about 1 in 1000 cases, the disease can be fatal. Prior to 2013, chikungunya virus cases and outbreaks had only been identified in countries in Africa, Asia and the Indian and Pacific Oceans. In late 2013, the first local transmission of chikungunya virus in the Americas was identified in Caribbean countries and territories and has been spreading since.

Zika is also caused by a virus with that name. Until recently, it wasn't thought to cause any significant problems in humans – in 80% of infections no symptoms at all occur while in the remaining 20% mild flu-like symptoms are present. In 2015, a large number of babies were born in Brazil with very severe malformations, especially of the brain and skull. This was eventually traced to zika infection of the mother during pregnancy.

For all three of these infections, climate change represents a significantly increased risk for humans. Already dengue infections have markedly increased in India and recent infections were found as far north as Paris. Increased temperature, increased humidity, wind, rainfall, flood, and drought are pushing mosquitoes to move beyond their normal range, especially during storms or hurricanes.

Shipping and air travel transport mosquitoes and their eggs to new areas. Infected humans travel to areas where local mosquitoes might then pick up and transmit the viruses. Children, pregnant women and the elderly are most at risk from infections, often exacerbated by the presence of unplanned urban housing such as slums, water stressed areas (leading people to store water in uncovered containers that act as mosquito breeding grounds) and food insecure zones where sanitation can be scarce.

Cholera – This is caused by a bacterium called *Vibrio cholerae*. It lives in water but when taken in by humans after drinking infected water attaches to the gut wall and prevents further absorption of water by the gut. This causes severe diarrhoea and if not quickly treated means the sufferer dies very quickly and unpleasantly.

The diarrhoea produced contains large amounts of the bacteria and if this then reaches the water source more infections will ensue. Cholera sounds like a disease of the past and by far the greatest risk factor is poor sanitation (i.e. drinking water and sewage in the same water source) which in most countries has been dealt with. In fact, cholera has never gone away and cases are now on the rise. In 2017, an estimated cumulative total of 1.2 million cases, including over 5,600 deaths, were reported to the World Health Organization. Each year an estimated 1.3 to 4 million people get cholera and 21,000 to 143,000 die from it. The African continent has the highest case fatality rates.

Climate change will increase these numbers even further. Disruptions to water supply – flood or drought – increase the risk of cholera and typhoons and earthquakes have been shown to leave a legacy of cholera secondary to water supply disruption. Poor sanitation from crowded human communities is a major risk. Ocean warming and heavy precipitation, which reduces coastal water salinity, appears to provide fertile conditions for *Vibrio cholerae*. According to the WHO, 30 countries reported outbreaks in 2022, 50% more than the previous years' average. Many of those outbreaks were related to tropical cyclones and their ensuing displacement of people. African countries reported 26,000 cases and 660 deaths in the first four weeks of 2023 compared to nearly 80,000 cases and 1,863 deaths during the whole of 2022. The last cholera outbreak in Venice was a little over 100 years ago (1911) but it will not be 100 years until the next one, probably.

Schistosomiasis – also known as Bilharzia, is caused by a parasitic worm. The larval form of this parasite is released from its host – freshwater snails – and can then penetrate the skin of humans who are using the water. The larvae hatch in the body and the adult worms live in the sufferer's blood vessels. They release eggs which are passed out in urine and faeces to allow another cycle of infection.

Schistosomiasis is almost exclusively found in tropical and subtropical conditions, with 90% of cases in Africa. The WHO estimates that in 2021 over 250 million people were infected in 78 countries. It mostly affects poor and rural communities, particularly agricultural and fishing populations. Women doing domestic chores in infested water, such as washing clothes, are also at risk. Inadequate hygiene and contact with infected water make children especially vulnerable to infection.

Symptoms of schistosomiasis are caused mainly by the body's reaction to the worms' eggs. Intestinal schistosomiasis causes abdominal pain, diarrhoea and blood loss. Liver damage can develop and if it continues can be fatal. Urogenital schistosomiasis causes blood in urine, kidney damage and in the longer term bladder scarring can occur. In some cases the latter develops into bladder cancer. Urogenital schistosomiasis may present with painful, bleeding genital lesions and is associated with fertility loss.

Climate change will have a more complex effect on schistosomiasis than some other infections. Raised temperatures will initially enhance the snail populations, making infections more likely but as temperatures continue to rise this will become detrimental to the snails. However, they will then be able to migrate to previously colder areas that are warming – so we will see the disease slowly spreading out of Africa and South Asia. Water quality issues will significantly increase the risk with poor sanitation arising from water shortages and greater concentrations of populations that have moved from hotter regions.

Amoebiasis – this is caused by a protozoa called *Entamoeba histolytica*. The active form of this amoeba lives in humans and other animals. The cysts it produces (similar to the eggs from worms that hatch into adults when conditions are optimal) are excreted in human faeces and can live in water, soils and damp food for some considerable time. Infection then occurs by contact with an infected source and subsequent swallowing of the cysts which then mature into the adult form.

This adult form resides in the gut, often without symptoms but in 10-20% causes dysentery (amoebic dysentery) which is fever, abdominal pain and severe diarrhoea often containing blood. Sometimes the amoeba can leave the gut and invade the liver causing inflammation and liver function problems. Rarely, but fatally, they can reach the brain. *E. histolytica* is estimated to infect about 35-50 million people worldwide and is estimated to kill more than 55,000 people each year mainly through dehydration from the severe diarrhoea – especially children.

Although the amoeba is found worldwide, it thrives in tropical and sub-tropical temperatures. Thus, once again, higher global temperatures will encourage it both to increase its numbers and range – infections have already been reported in the USA. Poor sanitation and hand hygiene are common reasons for infection, thus floods, droughts and migrating populations are all risk factors. As with so many catastrophic events such as hurricanes, earthquakes and wildfires, disruption to the water supply is an early result and amoebiasis is another threat that follows shortly after.

Plague – Yes, Plague or the Black Death or Bubonic Plague. Most of us think this disease disappeared in medieval times. It didn't and still causes outbreaks and deaths but just in places we don't really hear of. Under the influence of climate change though, we will be hearing much more about it.

Plague is caused by a bacteria called *Yersinia Pestis*. This bacteria lives in fleas and the fleas live on rats (also some squirrels, mice and rabbits). The fleas feed on the rat's blood but if there is a shortage of rats they will find other animals to feed from – during which they pass on the bacterium. Infection can also occur between humans when an infected person coughs, expels some bacteria, which are then breathed in by others.

The incubation period of bubonic plague is usually 2 to 8 days. Patients develop fever, headache, chills, and weakness and one or more swollen, painful lymph nodes (these are called buboes – hence the name). This form usually results from the bite of an infected flea and the bacteria multiply in a lymph node near where the bacteria entered the human body. If the patient is not treated with the appropriate antibiotics, the bacteria can spread into the bloodstream and then can travel anywhere in the body – this is called septicaemic plague. Sufferers develop fever, chills, extreme weakness, abdominal pain and bleeding into the skin and other organs. Skin and other tissues may turn black and die, especially on fingers, toes, and the nose. The mortality rate at this stage is extremely high.

Bacteria breathed directly into the lungs gives rise to pneumonic plague. This produces mainly lung symptoms – shortness of breath, cough, sputum and can lead to complete lung failure. Mortality rate is again very high.

At present, plague exists in all regions except Europe. Notable plague outbreaks have occurred in several Asian, African, and South American countries in the past 10 years. Climate change will alter where plague outbreaks occur – mainly through changes in the behaviour of fleas. It has been long suggested that seasonal variations in temperature and humidity are responsible for the seasonal patterns of human plague incidence. Human plague outbreaks in several African countries are less frequent when the weather was too hot (greater than 27°C) or cold (less than 15°C). Subsequent studies have shown an increased plague incidence in Vietnam during the hot, dry season when it is followed by a period of high seasonal rainfall (exactly what occurs with climate change). These ideal conditions for fleas/plague are slowly making their way north.

<u>West Nile Virus (WNV)</u> – which you may never have heard of but is actually the leading cause of mosquito-borne disease in the continental USA. WNV was first detected in the West Nile district of Uganda in 1937, and the first large outbreak in Europe occurred in Romania in 1996. Cases have been identified in several countries across Europe including France, Italy, Portugal and Spain. In 1999, WNV spread to North America and has subsequently been detected in all states except Alaska and Hawaii. Between 1999 and 2016, there were more than 45,000 cases and over 2,000 deaths. The main reservoirs are birds and when the mosquitos bite the birds they also get the West Nile virus which they pass on to the humans they later bite. Rats and squirrels can also be vectors and show again how interconnected we are with all the world's creatures. Changing their habitats and numbers can create threats to us that we cannot foresee.

Most infections in humans are asymptomatic (80%). In those who develop symptoms, there is a mild flu-like febrile illness, sometimes called West Nile Fever. A small proportion (less than 1%) of those infected will develop more severe disease – usually encephalitis (brain inflammation) or meningitis. People over the age of 50 are more likely to develop this severe disease and about 10% of neurological infections are fatal.

Climate change has already expanded the reach of the carrying mosquitos – hence the increase in cases in Europe and the USA. The main carrying mosquito was detected in the UK in 2010 although so far there have not been any cases of infection.

<u>Chagas disease</u> – Chagas disease or American Trypanosomiasis, is caused by the protozoan parasite *Trypanosoma cruzi*, which is transmitted to animals and people by insect bites. It is found only in the Americas (mainly in rural areas of Latin America). It is estimated that as many as eight million people in Mexico, Central America, and South America have Chagas disease, most of whom do not know they are infected. If untreated, infection is lifelong and can be life threatening.

Chagas disease has an acute phase and chronic phase. The acute phase presents with a non-specific flu-like illness and sometimes large boils (chagomas) at the site of the insect bite. This phase usually passes but in some cases heart and brain infections occur which can be life-threatening. The chronic form may not occur for years or even decades after initial infection. This may include heart or gut involvement as the muscle of the wall is invaded and can then be fatal.

Climate change will see Chagas disease spread into the Southern USA and then further northwards. If temperatures exceed 30°C and humidity does not increase sufficiently, the bugs increase their feeding rate to avoid dehydration. Similarly, it has been shown this rise in temperature makes the vector insects develop shorter life cycles and so higher population densities. High temperatures also accelerate development of the trypanosome itself and increases its dispersal in the locality.

Escherichia coli – or *E.coli* for short is a bacteria we have all come into contact with as many billions are living in your gut at this precise minute. There are a wide variety of species of *E.Coli*, some of which cause diarrhoeal diseases when swallowed. *E.coli* diarrhoea is nasty but not life threatening, unless in the very young and old or very unlucky. It is the commonest cause of 'food poisoning' in the world.

A study in 2016 found an 8% increase in the incidence of diarrhoea causing *E.coli* for each 1°C increase in mean monthly temperature. Wherever you live in the world, you live alongside *E.Coli* and its vicious siblings who are waiting for the temperature to rise.

Food poisoning, which is caused by bacteria we swallow when eating infected food, is going to increase with climate change. *E.Coli* is the most likely organism but *Salmonella* bacteria will also be more prevalent. Higher temperatures, more intense rainfall, flood and drought all make contamination of fresh produce much more likely. Higher temperatures have also been shown to increase *Salmonella* and *Campylobacter* in chicken flocks. Seafood is a prime target for food poisoning organisms, and they will flourish further in warmer seas, with higher nutrient levels and the raised sea levels that climate change brings.

Other diarrhoeal disease-causing organisms will also relish our warmer world. A study from China commented that for Cryptosporidium: 'Under future climate change, non-suitable habitats for Cryptosporidium will shrink, while highly suitable habitats will expand significantly'.

A different study said this about Giardiasis: 'Besides temperature, other factors that can increase the risk of giardiasis and directly related with climate change are precipitation/humidity and wind/dust'.

A stark and very messy future awaits us.

Candida – Candida is a species of fungus found all over the world and in all environments. It lives in and around humans and usually does not cause us health problems. However, if our defences are compromised by severe illness, especially if associated with intensive care unit stays and prolonged use of antibiotics, *Candida auris* can be deadly.

C. auris is multidrug resistant and therefore very hard to treat. It is a serious worldwide problem in intensive care units and other hospital environments. Climate change seems to be amplifying this problem with increased outbreaks documented in the USA, Canada, Panama, Colombia, Chile and Venezuela.

C. auris was originally found in the soil of ecosystems with some salinity, such as wetlands. Higher average temperatures resulting from global warming seem to have acted as a selective pressure on *C. auris*, favouring strains adapted to salinity and higher temperatures. These environmental conditions are similar to the conditions found in the human body, therefore creating increasingly favourable conditions for *C. auris* to infect humans. As the global temperature rises, it gets nearer and nearer to that of human body temperature, meaning that organisms that thrive in these temperatures can more easily cause us disease. **We are creating the most fertile of conditions for these old enemies to successfully attack.**

Rising temperatures are also allowing other disease-causing fungi to spread into new areas that previously were too cold for them to survive. For example, Valley Fever – caused by a fungus that lives in the soil in hot and dry areas – has already spread into the Pacific Northwest of the USA. This fungus can cause severe infections and death and is often misdiagnosed partly because of its previous rarity. Once again it appears that as the difference between environmental temperatures and human body temperatures narrows, this significantly aids the infection that causes Valley Fever.

Climate change also of course increases the risk of natural disasters and flooding, which in turn increase the risk for these moulds to grow in people's homes. Fungi that cause diseases are likely to become more and more of a problem for all the reasons listed above. These infections are notoriously difficult to eradicate and will, once again, put significant pressure on healthcare systems worldwide.

Rabies – Rabies is a much-feared infection, caused by a virus transferred by a bite. It is 100% fatal to the unvaccinated and can infect all mammals. In rabies-endemic countries, nearly 99% of the cases in humans are caused by dog bites. The World Health Organization estimates that between 59,000 and 60,000 human deaths are caused by rabies each year, with South Asian countries responsible for approximately 45% of these deaths. Around 59% of global rabies deaths are recorded in Asia, followed by 36% in Africa. Most of these nations fall into the category of low- and middle-income countries. The remaining 5% of human deaths occur in other low- and middle-income countries with relatively few cases reported from high-income countries.

Viruses are not affected by climate change or environmental temperature so why would rabies cases increase? The answer lies again in the vector that carries the disease, and in this case it is bats. Bats carry the rabies virus and pass it on through their bites when they feed on mammals. Climate change has allowed the bats to flourish in what previously would be too cold environments, at higher altitudes and for longer as winters become less cold. Because more bats are active and able to travel greater distances in warmer weather, rabies is more likely to spread. Thus as the temperature goes up, so has the number of reported cases of rabies. As we continue to warm, this number will continue to increase as the spread of the disease northwards.

Trachoma - Is a horrible eye disease caused by *Chlamydia trachomatis* and is the world's leading infectious cause of blindness. It is thought to be endemic in 51 countries, primarily in Sub-Saharan Africa. It is estimated that 232 million people living in trachoma-endemic areas are at risk of the condition. The infection causes inflammation and then scarring of the eyelids, which in turn causes the eyelashes to scrape painfully across the front of the eye – the cornea. As this damage continues the cornea becomes more and more scarred until, like a broken car windscreen, it becomes opaque. Of the over 1.8 million who are visually impaired due to trichiasis over 500,000 people are irreversibly blind.

Chlamydia trachomatis is passed on directly by touching the eye with an infected finger or towel or clothes. It can also be passed on by flies moving from one person's face to another. Poor water quality and hand hygiene are major risk factors for passing the infection on as well as providing breeding grounds for the flies.

Climate change is responsible for an increase in cases worldwide. The higher temperatures have been shown to increase the number of flies as well as their range. Floods and droughts disrupt the water supply and make hand hygiene much more difficult. It has been suggested that in Spain, where trachoma was eradicated generations ago, it could reappear in the near future.

Climate change is going to have a significant impact upon other blinding diseases also. For example, river blindness (Onchocerciases) which we mentioned earlier, has all the elements of a disease that will be accelerated by climate change.

Toxoplasmosis, which is a parasitic infection that attacks the retina of the eye, will be more commonplace. Worm infestations such as the *Loa Loa* worm (Loiasis) will become more common – both because the environmental conditions will increasingly suit it and because vector numbers will increase. This infection occurs after the bite of a fly and the worm injected can live and grow inside the surface skin of the eye (the conjunctiva). If you want to see a video of it being removed from the eye surface there are plenty to see on *YouTube*.

I wonder sometimes if the *Loa Loa* worm could be used as the 'poster child' of climate change. Videos of it being removed from the eye might give us humans the jolt we seem to need to stop wrecking the Earth's climate.

Anthrax – this again seems a disease of the distant past – a word rather than a threat. Anthrax is caused by a bacteria called *Bacillus anthracus*, and usually passed on after direct exposure from infected cattle. Worldwide, there are few cases per year, but this might be changing.

Global warming has resulted in the thawing of the frozen soil (permafrost) in Siberia and the consequent exposure of humans and animals to *Bacillus anthracis* spores that were safely locked in this ice for generations. In 2016, an outbreak of anthrax that killed thousands of reindeer and infected dozens of humans occurred in the Yamal peninsula, Northwest Siberia. This was the first outbreak for over 70 years. The trigger of the outbreak was shown to be the activation of spores from permafrost thaw that was accelerated during the summer heatwave as well as by abnormal rainfall.

This thawing of previously permanently frozen soil has the capacity to unleash a number of diseases. Many viruses and bacteria and fungal spores can survive being frozen and still awaken to cause infection. Not only is permafrost a risk for this but other previously ice-locked areas in the Arctic and Antarctic as well as the alpine glaciers present a risk. It might be too late by the time we find out how many and which infectious organisms are awaiting defrosting. It is possible that we have never been exposed to some of them and therefore have no immunity.

Leprosy (Hansen's Disease) – again a disease you might think of only in historical terms, but it still exists – just in places you don't really know about. It is caused by a bacteria called *Mycobacterium leprae* which damages the nerves of the body or the lungs causing ongoing damage to both. The sufferer cannot feel pain so damage their limbs without realising and this eventually causes loss of fingers, toes and limbs.

It is mainly caught by close proximity to someone with the bacteria, but a study in Ethiopia found that a transmission of leprosy can be changed with environmental factors. There is some evidence of a correlation between heat and humidity and the prevalence of leprosy. Once again, we are creating the ideal climatic conditions that might promote yet another disease that we hoped had been eradicated.

Any increase in leprosy will not just be restricted to Africa. Between 2002 and 2014, Florida reported around 10 cases a year, with a low of two cases in 2005 and a high of 12 cases in 2010. The numbers then rose to a high of 29 in 2015, however, and have stayed higher, with 27 cases reported in 2020. With 159 new cases reported across the US this is an increase from around 6% in 2014.

Tuberculosis (TB) – this seems to be a very Victorian disease but there are plenty of cases worldwide. It is caused by a bacteria called *Mycobacterium tuberculosis* which is usually passed from person to person especially when in close and prolonged proximity to an infected person.

There is no evidence that the bacteria itself is promoted by climate change nor if there is a vector that is enhanced. But climate change will increase TB in a different way. We have discussed already the effects of climate change upon migration and subsequent spreading of diseases and most importantly with the crowding this migration brings. It is this crowding that will allow the tuberculosis bacteria to spread from person to person.

It is easy to forget that some of the more indirect effects of climate change are just as potentially threatening to our health as the more direct ones.

And more and more...

Climate change is so all encompassing that it can affect all organ systems and regions of the body. I have discussed the major diseases – life threatening and life quality reducing – but I could have added many more. From risks to hearing and balance to feet infections to skin ageing, the evidence exists for a direct effect from climate change. Every health practitioner will be affected by the scale of the diseases they will face. Every healthcare system will be swamped by demand and inevitably some will collapse under the weight.

Added to this, the mechanisms of climate change and diseases can be calculated and postulated, but we will never be able to predict fully the new diseases we might face in our new environment. Climate scientists have warned of this for some time – we are pushing the climate to a place it has not been for millions of years and long before human history. We therefore do not have the experience to know what we may unleash.

From directly increasing the potency and range of some of the organisms themselves, to increasing the vector numbers, to reducing human resistance to infection, to declining efficacy of treatment, all these make pretty much any infectious disease more likely. They are a potential horror show of suffering and death and, as in all good horror films, they will appear in different guises and when least expected.

Chapter Summary

- Very few organ systems won't be adversely affected by the changes to our climate.

- Diseases ranging from fungal foot infections to strokes and heart attacks and cancer will be more likely as our environment warms.
- Certain individuals will suffer these affects disproportionately, such as young children, the elderly, outdoor workers and the poor.
- The rise in infectious diseases present one of the most serious threats to humans from global warming.
- Not only will there be an increase in the number of infections but they will spread to other parts of the world where they were previously unheard of.

THE END

This book is not, as many climate warning books are, a book about solutions to the problems it presents. The solutions that remain possible (we are too late for some) to prevent our planet warming further are inconvenient, although obvious. The thing is, unfortunately, that these solutions require sacrifices and none of us are very good at giving up what we want.

If you feel frightened after reading about the threats that face us (as I am) then try to turn this fear into something positive. Adapt your own life to think of the climate first, and tell everyone who will listen (and especially those who don't want to listen) about the dangers we face. All we humans are responsible for this mess, and so we are all responsible for clearing it up. You may feel you are too insignificant to make things better but remember you are not too insignificant to make things worse.

I won't waste words by giving pages of suggestions for change, saying what all articles feel they have to end with – the word 'hope'. The problem with using the word 'hope' is that it is too easily subverted. If we talk about hope for the future planet in the light of the current trajectory of global warming, what do we mean? Do we mean we hope it doesn't happen? Do we mean that we hope God intervenes? Do I hope that it just isn't too bad in my lifetime? Or do we use it in the context it is usually meant – that we have hope but only if we mend our ways now? It depends of course on where you stand on the issue. So let's not use words like hope, but just understand that we really have no choice and that we need to make big changes **now.**

All humans, by their very presence on this planet, produce CO_2 and are affected by CO_2. Some of us only produce very small amounts by simply breathing out. Others produce huge amounts from business flights and constant cruises. The more of us there are and the more we emit, the worse the situation will get. We humans are the only organisms who have ever inhabited this planet who know what our fate will be if we continue our current destructive behaviour. The dinosaurs at least had the excuse of ignorance.

If you don't care for the beauty of this planet, if you don't love the look of trees and flowers and tigers and angel fish, then do you not care about yourself? Do you not care about those you love? If you don't, I am surprised you are reading this book. Whichever category you are in, please take note of the obvious warnings in this book. We are all sitting on the same branch and we all are using the same saw.

When researching for this book the phrase '...will worsen as climate change intensifies' was written so much on so many topics that it became meaningless in my mind. But that makes me the same as anyone else who crosses their fingers while looking the other way. We **know** what is coming as it is spelt out over and over again in scientific studies. We can see the speeding train coming towards us, but instead of moving tracks we just hope it will stop before it hits us.

No age group is immune from all of this. Older citizens might feel that the worst effect of climate change will happen after they die. That only works if you don't care about future generations of humans, and that would inevitably include your own offspring. But it is intrinsically false as we have seen. The elderly are disproportionately affected by the hot days, the sleepless nights, the increased crime and the polluted streets that are with us **now**. As is all too clear, climate change is affecting all facets of our life, and all facets of our life are affecting climate change.

If there is one underlying message of this book, it is that even small changes in complex systems can have unpredictable results. The planet is a metaphor for our bodies, and our bodies are a metaphor for our planet. We continue to change our planet's climate in all sorts of big and small ways, but we cannot **predict** the outcomes of these changes and we can only know once we have lived through them. Similarly, if you abuse your health – even in a minor way – you can't predict the long term outcome of that action and only in the future will you know if there are consequences.

This, of course, is not accidental. Evolution has finely tuned us to the planet. The planet itself, unlike its inhabitants, does not evolve, but changes depending upon the stresses and strains put upon it. So logic would tell us that we shouldn't disrupt this fine tuning – we can bend the planet to our will, but we can't know how the changes we create will disrupt our own relationship with it and therefore our existence upon it.

I must finish with a phrase that I used in the introduction, and which I wish we all would bear in mind. It is perhaps becoming a cliché, but is no less useful for that. **The Earth will survive the age of humans but the humans probably won't.**

If you have read all this book, then now you will know why.

The Hummingbird

One day a terrible fire broke out in the forest, and the huge woodland was suddenly engulfed by a raging wildfire. Frightened, all the animals fled their homes and ran out of the forest. As they came to the edge of a stream, they stopped to watch the fire and felt very discouraged and powerless.

They were all bemoaning the destruction of their homes. Every one of them thought there was nothing they could do about the fire – except for one little hummingbird. It swooped into the stream and picked up a few drops of water and went into the forest and put them on the fire. Then it went back to the stream and did it again, and it kept going back, again and again and again.

The other animals watched in disbelief. Some tried to discourage the hummingbird with comments like, 'Don't bother, it is too much, you are too little, you can carry only a drop, you can't put out this fire.'

One of the bigger animals shouted to the hummingbird in a mocking voice, 'What do you think you are doing?' And the hummingbird, without wasting time or losing a beat, looked back and said, 'I am doing what I can'.

Probably a folktale from the Quechua people of Peru

References and Notes

These are the main references and sources for each chapter. Some papers, articles and reports may have been used in multiple chapters, but I will only reference when mentioned for the first time.

For more general reading please see the Further Reading section.

Chapter 1   Is the Earth getting warmer (and why)?

Global warming and its causes

Santer, Benjamin D. *et al*. (16 Sept 2013) Human and natural influences on the changing thermal structure of the atmosphere. *Proc Natl Acad Sci.* 110 (43) 17235-17240.
https://www.pnas.org/doi/full/10.1073/pnas.1305332110

The Royal Society, London. (Mar 2020) *A Short Guide to Climate Science.* www.royalsociety.org. This is a short but excellent summary of climate change science from the prestigious Royal Society:
https://royalsociety.org/-/media/policy/projects/climate-evidence-causes/climate-change-q-and-a.pdf

For more greater detail, facts and figures there is *Climate Change Evidence and Causes* (update 2020). This is a joint document from the US National Academy of Sciences and the UK Royal Society again:
https://royalsociety.org/-/media/Royal_Society_Content/policy/projects/climate-evidence-causes/climate-change-evidence-causes.pdf

Probably the most extensive resource for all aspects of climate change is from the Intergovernmental Panel on Climate Change (IPCC). Their various reports give huge amounts of information, data and evidence for global warming:
https://www.ipcc.ch/report/ar6/wg2/

NASA has produced its own measurement and summaries about the scale of the temperature rise and its cause:
https://climate.nasa.gov/evidence/

The hottest summer in human history
https://www.theguardian.com/environment/ng-interactive/2023/sep/29/the-hottest-summer-in-human-history-a-visual-timeline

Rises in ocean temperatures
https://www.epa.gov/climate-indicators/climate-change-indicators-sea-surface-temperature#:~:text=As%20the%20oceans%20absorb%20more,marine%20ecosystems%20in%20several%20ways

Ice loss from Greenland and Antarctica
https://www.esa.int/Applications/Observing_the_Earth/FutureEO/CryoSat/Ice_loss_from_Greenland_and_Antarctica_hits_new_record

To see a map of the how climate change affects weather all over the world look at:
https://www.carbonbrief.org/mapped-how-climate-change-affects-extreme-weather-around-the-world/

REFERENCES AND NOTES

For more about the history of the discovery of the Greenhouse Effect see:
https://www.rigb.org/explore-science/explore/blog/who-discovered-greenhouse-effect

For more detail upon the deadly effect of feedback loops on the crisis this is a good summary:
https://www.climaterealityproject.org/blog/how-feedback-loops-are-making-climate-crisis-worse

Finally some of the information in this chapter has come from Mark Lynas's excellent book *Our Final Warning. Six Degrees of Climate Emergency.* More information is in the Further Reading Section at the end of the book.

Chapter 2 What happens to Nature when the temperature rises (and why that's important to humans)

The IPCC remains a mainstay of climate information https://www.ipcc.ch/report/ar6/wg2/
For more ocean related IPCC information
https://www.ipcc.ch/srocc/

Drought and climate change
https://www.c2es.org/content/drought-and-climate-change/#:~:text=How%20climate%20change%20contributes%20to,would%20be%20in%20cooler%20conditions

United Nations Environment Programme on Wildfires
https://www.unep.org/resources/report/spreading-wildfire-rising-threat-extraordinary-landscape-fires?gclid=EAIaIQobChMIkuPkz8_HgQMVZItQBh0sNwgeEAAYASAAEgKlH_D_BwE

The effects of wildfires
https://www.nature.com/articles/s41577-022-00776-3

The UK National Meteorological Service (the Met Office) information upon extreme weather events
https://www.metoffice.gov.uk/research/climate/understanding-climate/uk-and-global-extreme-events-heavy-rainfall-and-floods\

UN climate action and the oceans
https://www.un.org/en/climatechange/science/climate-issues/ocean

Coastal threats
He,Q. & Silliman,B. (October 2019) Climate Change, Human Impacts and the Coastal Ecosystems in the Anthropocene. *Current Biology* 29, 10121-1035.
https://www.cell.com/current-biology/fulltext/S0960-9822(19)31092-9?_returnURL=https%3A%2F%2Flinkinghub.elsevier.com%2Fretrieve%2F-pii%2FS0960982219310929%3Fshowall%3Dtrue

NASA data and ocean warming
https://climate.nasa.gov/vital-signs/ocean-warming

The effects upon sharks
https://www.sciencetimes.com/articles/45257/20230805/hotter-oceans-fueling-aggression-sharks-scientists-warn-record-temperatures-reached.htm

Effect of the ocean on the land
https://oceanexplorer.noaa.gov/facts/climate.html

Defining marine heatwaves
https://www.sciencedirect.com/science/article/abs/pii/S0079661116000057

Stalling of hurricanes
https://www.nature.com/articles/s41612-019-0074-8

Climate change and plants
https://news.climate.columbia.edu/2022/01/27/how-climate-change-will-affect-plants/

Pequeno *et al.* (2021) Climate impact and adaption to heat and drought stress of regional and global wheat production. *Environmental Res Lett.* 16: 1-17.
https://iopscience.iop.org/article/10.1088/1748-9326/abd970/pdf

The effect of climate change on insect populations
https://www.theguardian.com/environment/2019/feb/10/plummeting-insect-numbers-threaten-collapse-of-nature

Climate change and animal extinction threats
https://www.nature.com/articles/s41558-022-01490-7

Climate change and microorganisms
Cavicchioli *et al.* (18 June 2019) Scientists warning to humanity: microorganisms and climate change. *Nature Reviews Microbiology.* Vol 17, 569-586, issue date September 2019.
https://www.nature.com/articles/s41579-019-0222-5

Climate change, earthquakes and volcanoes
https://www.theguardian.com/world/2016/oct/16/climate-change-triggers-earthquakes-tsunamis-volcanoes

Chapter 3   What happens to humans as the temperature rises?

For further information on body temperature regulation and measurement see:
https://www.ncbi.nlm.nih.gov/pmc/articles/PMC8535559/

The World Meteorological Organization is a rich source of data:
https://library.wmo.int/#.WOOzUnew1Bw
The Climate Signals website is also a rich source of data and explanations. The heat and heatwaves sections are here:
https://www.climatesignals.org/climate-signals/extreme-heat-and-heat-waves

WHO and heatwave data:
https://www.who.int/health-topics/heatwaves#tab=tab_1

For a review of heat related health impacts:

Gauer, R. & Meyers, BK. (April 2019) Heat-Related Illnesses. *American Family Physician*. Vol 99(8), 482-489.
https://www.aafp.org/pubs/afp/issues/2019/0415/p482.html.

Edi *et al.* (Aug 2021) Hot weather and heat extremes: health risks. *The Lancet*. Volume 398. 698-708.
https://pubmed.ncbi.nlm.nih.gov/34419205/

Heatwaves are becoming more common
https://www.who.int/health-topics/heatwaves#tab=tab_1
https://phys.org/news/2020-05-potentially-fatal-combinations-humidity-emerging.html

Chapter 4   Threat by Threat

The WHO and World Bank climate and health
https://www.who.int/news/item/18-09-2023-leaders-spotlight-the-critical-intersection-between-health-and-climate-ahead-of-cop-s-first-ever-health-day
Climate change and earthquakes
https://worldcrunch.com/green/climat-change-earthquakes-connection

REFERENCES AND NOTES

Health effects of climate change
https://bmjopen.bmj.com/content/bmjopen/11/6/e046333.full.pdf

https://www.ncbi.nlm.nih.gov/pmc/articles/PMC6210172/

WHO earthquake statistics
https://www.who.int/health-topics/earthquakes#tab=tab_1

Climate change and volcanic eruptions
https://www.preventionweb.net/news/volcano-erupting-again-iceland-climate-change-causing-more-eruptions

Volcano mortality
Brown, S.K., Jenkins, S.F., Sparks, R.S.J. *et al.* (2017) Volcanic fatalities database: analysis of volcanic threat with distance and victim classification. *J Appl. Volcanol.* 6, 15.
https://doi.org/10.1186/s13617-017-0067-4

Wind damage
Marchigiani *et al.* (2013 Apr-Jun) Wind disasters: A comprehensive review of current management strategies. *Int J Crit Illn Inj Sci.*; 3(2): 130–142. doi: 10.4103/2229-5151.114273.
https://www.ncbi.nlm.nih.gov/pmc/articles/PMC3743338/

Mortality risks from cyclones
https://www.thelancet.com/journals/lanplh/article/PIIS2542-5196(23)00143-2/fulltext?dgcid=raven_jbs_etoc_feature_lanplh

UN Environment Programme Wildfire Report
https://www.unep.org/resources/report/spreading-wildfire-rising-threat-extraordinary-landscape-fires?gclid=CjwKCAjwloynBhBbEiwAGY25dGNSn1RhLeMs-s3eixLwp5VDnw8jnEuANabHqEuLO90mAiWpN1hEJBoCVq8QAvD_BwE

Estimating a wildfire speed
https://www.iawfonline.org/article/evaluating-the-10-wind-speed-rule-of-thumb-for-estimating-a-wildfires-forward-spread-rate/

Prevalence and effects of wildfires
Akdis, C.A., Nadeau, K.C. (2022) Human and planetary health on fire. *Nat Rev Immunol* 22, 651–652.
https://doi.org/10.1038/s41577-022-00776-3

Economics of wildfires
https://www.nist.gov/publications/costs-and-losses-wildfires#:~:text=The%20annualized%20economic%20burden%20from,%2463.5%20billion%20to%20%24285.0%20billion

Drowning risk and climate change review
https://injuryprevention.bmj.com/content/28/2/185

Increased winter drownings
https://journals.plos.org/plosone/article?id=10.1371/journal.pone.0241222

World Meteorological Organization and storm surges
https://public-old.wmo.int/en/our-mandate/focus-areas/natural-hazards-and-disaster-risk-reduction/storm-surge

The US Natural Resources Defence Council Report on flooding and climate change
https://www.nrdc.org/stories/flooding-and-climate-change-everything-you-need-know#facts

WMO and Drought
https://public-old.wmo.int/en/resources/world-meteorological-day/previous-world-meteorological-days/climate-and-water/drought

REFERENCES AND NOTES

Centres for Disease Control and Prevention (CDC), Health and Drought
https://www.cdc.gov/nceh/drought/implications.htm

The UN Sustainable Development Report 2023
https://dashboards.sdgindex.org

Royal Institution of Chartered Surveyors, climate change and coastal regions
https://ww3.rics.org/uk/en/journals/land-journal/how-does-climate-change-affect-coastal-regions-.html

Climate and glaciers
https://www.climate.gov/news-features/understanding-climate/climate-change-mountain-glaciers

Record breaking hailstones
https://www.scientificamerican.com/article/is-climate-change-causing-more-record-breaking-hail/

Cognitive dissonance
https://www.pbs.org/newshour/science/how-your-brain-stops-you-from-taking-climate-change-seriously

The World Food Programme – climate and famine
https://www.wfp.org/stories/act-now-climate-crisis-or-millions-more-will-be-pushed-hunger-and-famine

Desertification
https://www.nature.com/articles/s41467-020-17710-7

Climate change and fishing
https://www.gla.ac.uk/news/headline_966114_en.html#:~:text=Over%20the%20last%20century%2C%20global,to%20adapt%20to%20warmer%20conditions

WWF and seafood
https://www.worldwildlife.org/industries/sustainable-seafood#:~:text=More%20than%203%20billion%20people,to%20billions%20of%20people%20worldwide

Elevated CO_2 and crops
https://onlinelibrary.wiley.com/doi/10.1111/j.1365-2486.2007.01511.x

Algal blooms and health
https://www.thelancet.com/journals/ebiom/article/PIIS2352-3964(23)00169-X/fulltext

Microorganisms and climate change
https://www.nature.com/articles/s41579-019-0222-5

Air Pollution and antibiotic resistance
https://www.thelancet.com/journals/lanplh/article/PIIS2542-5196(23)00135-3/fulltext?dgcid=raven_jbs_etoc_email

Mora *et al.* (2022) Over Half of known human pathogenic disease can be aggravated by climate change. *Nature Climate Change* volume 12, pages 869–875.
https://www.nature.com/articles/s41558-022-01426-1

Dog bites
https://www.nature.com/articles/s41598-023-35115-6

Ant aggression and heat
https://www.sciencedirect.com/science/article/pii/S0048969722075453

Animal aggression
https://pubmed.ncbi.nlm.nih.gov/36716887/

UK Government Report on Air Pollution and Health
https://www.gov.uk/government/publications/health-matters-air-pollution/health-matters-air-pollution

Vicious cycles of heatwaves and air pollution
https://www.theguardian.com/environment/2023/sep/06/world-meteorologists-point-to-vicious-cycle-of-heatwaves-and-air-pollution

The Stern Report
https://webarchive.nationalarchives.gov.uk/ukgwa/20100407172811/
https:/www.hm-treasury.gov.uk/stern_review_report.htm

The Lancet Countdown Report in 2022 on health and climate change is a very detailed and comprehensive resource:
https://www.thelancet.com/article/S0140-6736(22)01540-9/fulltext

British Medical Association – Health and Poverty
https://www.bma.org.uk/media/2084/health-at-a-price-2017.pdf

War and climate
https://unfccc.int/blog/conflict-and-climate

Kuwait 30 years on
https://www.theguardian.com/environment/2021/dec/11/the-sound-of-roaring-fires-is-still-in-my-memory-30-years-on-from-kuwaits-oil-blazes

Mass migration and climate change
Clark-Ginsberg *et al.* (2023) Climate change-related mass migration requires health system resilience. *Environ Res Health* 2023:1;045004.
https://iopscience.iop.org/article/10.1088/2752-5309/ace5ca/pdf

Drowning risks and migration
https://injuryprevention.bmj.com/content/28/2/185#ref-61

Climate and crime
Peng & Zhan. (October 2022) Extreme climate and crime: Empirical evidence based on 129 prefecture-level cities in China. *Front.Ecol.Evol.*
https://www.frontiersin.org/articles/10.3389/fevo.2022.1028485/full

Heat and behaviour
Sinister & Cooper. (October 2004) Thermal Stress in the U.S.A.: effects on violence and on employee behaviour. *Stress and Health.*
https://onlinelibrary.wiley.com/doi/abs/10.1002/smi.1029

https://www.pnas.org/doi/10.1073/pnas.2204076119#sec-3

Anderson, CA. (2001) Heat and Violence. *Current Directions in Psychological Science.* 10 (1)
https://www2.psych.ubc.ca/~schaller/308Readings/Anderson2001.pdf

Weather and traffic accidents
https://analysisfunction.civilservice.gov.uk/wp-content/uploads/2017/01/Road-accidents.pdf

Resistance to clean air projects
https://www.dailymail.co.uk/news/article-12439979/Nine-10-ULEZ-cameras-vandalised-southeast-London.html

https://www.bbc.co.uk/news/uk-england-oxfordshire-63716090

Air conditioners and greenhouse gases
https://www.nrel.gov/news/press/2022/nrel-shows-impact-of-controlling-humidity-on-greenhouse-gas-emissions.html#:~:text=The%20researchers%20calculated%20air%20conditioning,million%20tons%20from%20removing%20humidity.

Journalists and heat
Zong & Zhou. (2012) Under the Weather: The Weather Effects on U.S. Newspaper Coverage of the 2008 Beijing Olympics. *Mass Communication and Society.* Vol 15: 559-577.
https://www.tandfonline.com/doi/abs/10.1080/15205436.2012.677091

Hurricane Katrina
https://www.urban.org/sites/default/files/publication/51061/900929-initial-health-policy-responses-to-hurricane-katrina-and-possible-next-steps.pdf

Rudowitz, R., Rowland, D., Shartzer, A. (2006) Health care in New Orleans before and after Hurricane Katrina. *Health Aff (Millwood)* 25:w393–406.
https://pubmed.ncbi.nlm.nih.gov/16940307/

Healthcare costs and climate change
https://www.nrdc.org/sites/default/files/costs-inaction-burden-health-report.pdf

Biodiversity and drug discovery
https://www.ncbi.nlm.nih.gov/pmc/articles/PMC5735771/#:~:text=This%20ongoing%20loss%20of%20biodiversity,every%20two%20years%20%5B4%5D

Healthcares' effect upon the climate
https://www.healthpolicypartnership.com/the-nexus-between-climate-change-and-healthcare/

https://noharm-global.org/sites/default/files/documents-files/5961/HealthCaresClimateFootprint_092319.pdf

https://www.managedhealthcareexecutive.com/view/healthcare-and-climate-change

Chapter 5   Disease by Disease

Ozone, temperature and skin disease
Silva, GS. & Rosenbach, M. (2021) Climate change and dermatology. *International Journal of Women's Dermatology*. 7(1): 3-7.
https://www.ncbi.nlm.nih.gov/pmc/articles/PMC7435281/
Neale, RE. et al. (2021) Environmental effects of stratospheric ozone depletion, UV radiation and interactions with climate change> UNEP Environmental Effects Assessment Panel, Update 2020. *Photochemical and Photobiological Sciences*. 20:1-67.
Climate change and skin cancer incidence
Rawlings Parker, E. (2021) The Influence of climate change on skin cancer incidence – A review of the evidence. *International Journal of Women's Dermatology*. 7(1): 17-27.
https://www.ncbi.nlm.nih.gov/pmc/articles/PMC7838246/

Fungal disease and climate change
https://www.cdc.gov/fungal/climate.html#:~:text=With%20shifting%20temperatures%2C%20fungi%20may,adapted%20to%20surviving%20in%20humans.&text=Heat%20may%20also%20cause%20other,of%20fungi%20to%20infect%20people.

Skin infections and climate change
https://dermnetnz.org/topics/climate-change#:~:text=or%20social%20isolation).-,Cutaneous%20infection%20associated%20with%20a%20warmer%20climate,Viral%20skin%20conditions.

Climate change and dermatology
Fathy, R. & Rosenbach, M. (2020) Climate Change and Inpatient Dermatology. *Current Dermatology Reports*. 9:201-209.
https://www.ncbi.nlm.nih.gov/pmc/articles/PMC7442546/

REFERENCES AND NOTES

Climate change, air pollution and health
https://www.worldbank.org/en/news/feature/2022/09/01/what-you-need-to-know-about-climate-change-and-air-pollution
Climate change and respiratory disease
https://err.ersjournals.com/content/23/132/161

https://breathe.ersjournals.com/content/breathe/19/2/220222.full.pdf

https://www.lung.org/clean-air/climate-change

Climate change and allergens
Beggs *et al.* (2023) Climate change, airborne allergens and three translational mitigation approaches. *The Lancet.* Vol 93:1-9.
https://www.ncbi.nlm.nih.gov/pmc/articles/PMC10363419/

Air Pollution and COPD
Hansel *et al.* (June 2016) The effects of air pollution and temperature on COPD. *COPD* 13(3):372-379.
https://www.ncbi.nlm.nih.gov/pmc/articles/PMC4878829/pdf/nihms-788046.pdf

Air pollution and lung disease
https://www.gov.uk/government/publications/health-matters-air-pollution/health-matters-air-pollution

Air pollution and lung cancer
https://www.esmo.org/newsroom/press-releases/scientists-discover-how-air-pollution-may-trigger-lung-cancer-in-never-smokers

COVID and climate change
https://www.hsph.harvard.edu/c-change/subtopics/coronavirus-and-climate-change/

Deaths from cardiovascular disease
https://world-heart-federation.org/news/deaths-from-cardiovascular-disease-surged-60-globally-over-the-last-30-years-report/#:~:text=Search%20for%3A%20Search-,Deaths%20from%20cardiovascular%20disease%20surged%2060%25%20globally,the%20last%2030%20years%3A%20Report&text=GENEVA%2C%2020%20May%202023%20-%20Deaths,World%20Heart%20Federation%20(WHF).

Heat and cardiovascular health
Liu *et al.* (2022) Heat exposure and cardiovascular health outcomes: a systematic review and meta-analysis. *The Lancet Planetary Health.* 6:484-95.
https://www.thelancet.com/action/showPdf?pii=S2542-5196%2822%2900117-6

Heatstroke and death
Hifumi *et al.* (2018) Heat Stroke. *J Intensive Care.* 6:30.
https://www.ncbi.nlm.nih.gov/pmc/articles/PMC5964884/pdf/40560_2018_Article_298.pdf

Climate change and cardiovascular disease
Jacobsen *et al.* (2022) Climate Change and the prevention of cardiovascular disease. *American Journal of Preventive Cardiology.* 12:1-13.
https://www.ncbi.nlm.nih.gov/pmc/articles/PMC9508346/

Climate change and cancer
Hiatt, RA. & Beyeler, N. (Nov 2020) Cancer and Climate Change. *Lancet Oncology.* 21(11):519-527. And Yu *et al.* (Jan 2023) Cancer and ongoing climate change: Who are the most affected? *ACS Enviro Au.* 3(1); 5-11.
https://www.ncbi.nlm.nih.gov/pmc/articles/PMC9853937/pdf/vg2c00012.pdf

Vineis *et al.* (2021) Climate change and cancer: converging policies. *Molecular Oncology.* 15;764-769.
https://www.ncbi.nlm.nih.gov/pmc/articles/PMC7931120/

Breast cancer and air pollution
https://www.earth.com/news/breast-cancer-incidence-linked-to-high-particulate-air-pollution/

Wildfires and cancer
Korsiak *et al.* (May 2022) Long-term exposure to wildfires and cancer incidence in Canada: a population-based observational cohort study. *Lancet Planet Health* 6(5):e400-e409.
https://pubmed.ncbi.nlm.nih.gov/35550079/

Nutrition and cancer
https://nutritionj.biomedcentral.com/articles/10.1186/1475-2891-3-19?ref=healthdecider

Health and secondhand smoke
https://www.cdc.gov/tobacco/secondhand-smoke/health.html

Toxicity of cigarette filters
https://link.springer.com/article/10.1186/s43591-022-00050-2

Climate change and the kidney
Johnson *et al.* (2019) Climate change and the kidney. *Annals of Nutrition and Metabolism.* 74(suppl 3);38-44.
https://karger.com/anm/article/74/Suppl.%203/38/42811/Climate-Change-and-the-Kidney

Kidney stones
Stamatelou *et al.* (Jul 2008) Time trends in reported prevalence of kidney stones in the United States. *Proc Natl Acad Sci USA.*105(28);9841-6.
https://www.kidney-international.org/action/showPdf?pii=S0085-2538%2815%2949072-6

Chronic kidney disease
Carney, EF. (Mar 2020) The impact of chronic kidney disease on global health. Nature Reviews Nephrology. 16;251.
https://www.nature.com/articles/s41581-020-0268-7

Soft drinks, heat, exercise and the kidney
Chapman, CL *et al.* (Mar 2019) Soft drink consumption during and following exercise in the heat elevates biomarkers of acute kidney injury. *Am J Physiol Regul Integr Comp Physiol.* 315(3);189-98.
https://pubmed.ncbi.nlm.nih.gov/30601706/

Chronic liver disease
Hirode *et al.* (Apr 2020) Trends in the Burden of Chronic Liver Disease Among Hospitalized US Adults. *JAMA* 3(4); e201997.
https://www.ncbi.nlm.nih.gov/pmc/articles/PMC7118516/

Climate change and liver disease
https://www.journal-of-hepatology.eu/action/showPdf?pii=S0168-8278%2822%2900118-0

Saad-Hussein *et al.* (2022) Role of Climate Change in Changing Hepatic Health Maps. *Current Environmental Health Reports.* 9; 299-314.

Algae and liver disease
https://www.cdc.gov/habs/be-aware-habs.html

https://www.publichealth.hscni.net/directorates/public-health/health-protection/severe-weather/health-risk-humans-blue-green-algae

Alcoholic liver disease
https://www.ncbi.nlm.nih.gov/pmc/articles/PMC9579565/pdf/41575_2022_Article_688.pdf

REFERENCES AND NOTES

Climate change and gut health
Sadeghi et al. (Apr 2023) Mini Review: The Impact of Climate Change on Gastrointestinal Health. *Middle East J Dig Dis.* v.15(2).
https://www.ncbi.nlm.nih.gov/pmc/articles/PMC10404088/pdf/mejdd-15-72.pdf

Cholera
https://www.who.int/emergencies/disease-outbreak-news/item/2023-DON437#:~:text=The%20average%20cholera%20CFR%20reported,trend%20for%202022%20and%202023

Climate change and gut bacteria
https://www.news-medical.net/news/20230426/Climate-change-linked-to-changes-in-gut-microbiota-and-aging-process.aspx#:~:text=By%20altering%20the%20respiration%20of,deplete%20the%20human%20gut%20microbiota.

Inflammatory bowel disease and air pollution
Kaplan et al. (Nov 2010) The Inflammatory Bowel Diseases and Ambient Air Pollution: A Novel Association. *Am J Gastroenterol.* 105(11): 2412–2419.
https://www.ncbi.nlm.nih.gov/pmc/articles/PMC3180712/

Acid reflux and weather
Ho Seok Seo et al. (Oct 21 2020) Relationship of meteorological factors and air pollutants with medical care utilization for gastroesophageal reflux disease in urban area. *World J Gastroenterol.* 26(39): 6074–6086.
https://www.ncbi.nlm.nih.gov/pmc/articles/PMC7584054/

Stomach ulcers and pollution
Shang-Shyue Tsai et al. (May 30 2019) Ambient Air Pollution and Hospital Admissions for Peptic Ulcers in Taipei: A Time-Stratified Case-Crossover Study. *Int J Environ Res Public Health.* 16(11):1916.
https://pubmed.ncbi.nlm.nih.gov/31151209/

Climate change and obesity. *Lancet review edition*
https://www.thelancet.com/commissions/global-syndemic

Obesity and climate change
https://earth.org/data_visualization/the-common-drivers-of-obesity-and-climate-change/#:~:text=The%20total%20impact%20of%20obesity,degradation%20and%20massive%20biodiversity%20loss.

Koch *et al.* (2021) Climate change and obesity. *Horm Metab Res.* 53;575-587.
https://www.ncbi.nlm.nih.gov/pmc/articles/PMC8440046/pdf/10-1055-a-1533-2861.pdf

Climate change and neurological disease
Luois *et al* (2023) Impacts of climate change and air pollution on neurologic health, disease and practice. *Neurology.* 100;474-483.
https://www.ncbi.nlm.nih.gov/pmc/articles/PMC9990849/

Climate change and dementia
Bongioanni *et al.* (Oct 2021) Climate change and neurodegenerative diseases. Environmental Research. Elsevier. *ScienceDirect.* 201:111511.
https://www.sciencedirect.com/science/article/abs/pii/S0013935121008057?via%3Dihub

Road traffic and dementia
https://www.alzheimers.org.uk/blog/busy-roads-dementia-risk-study-explained#:~:text=People%20who%20live%20near%20busy,of%20dementia%20in%20the%20past.
Climate change and the brain
https://www.news-medical.net/health/The-Impact-of-Climate-Change-on-Brain-Health.aspx

Body temperature regulation in diseases
Cramer *et al.* (Oct 1 2022) Human temperature regulation under heat stress in health, disease, and injury. *Physiol Rev.* 102(4): 1907–1989.
https://www.ncbi.nlm.nih.gov/pmc/articles/PMC9394784/

REFERENCES AND NOTES

Heat related dementia admissions
Gong *et al.* (Jan 15 2022) Current and future burdens of heat-related dementia hospital admissions in England. *Environ Int.* 159: 107027.
https://www.ncbi.nlm.nih.gov/pmc/articles/PMC8739554/#:~:text=A%204.5%25%20increase%20in%20risk,increase%20above%2017°C.&text=Emergency%20admissions%20may%20increase%20by,2040%20under%20high%20emissions%20scenario.&text=Identifying%20people%20with%20dementia%20as,population%20over%20summer%20is%20advocated

Air pollution and bone fractures
Prada *et al.* (2017) Association of air particulate pollution with bone loss over time and bone fracture risk: analysis of data from two independent studies. *Lancet Planet Health.* 1:e337-47.
https://www.thelancet.com/action/showPdf?pii=S2542-5196%2817%2930136-5

Air pollution and osteoporosis
https://www.publichealth.columbia.edu/news/air-pollution-speeds-bone-loss-osteoporosis#:~:text=Elevated%20levels%20of%20air%20pollutants,Mailman%20School%20of%20Public%20Health

Climate change and dental health
Patil, VS. (Jan-Mar 2023) Addressing the impact of the climate crisis on oral health. *International Journal of Preventive and Clinical Dental Research* 10(1):p 20-22.
https://journals.lww.com/inpc/fulltext/2023/10010/addressing_the_impact_of_the_climate_crisis_on.6.aspx#:~:text=affects%20oral%20health.-,INCREASED%20RISK%20OF%20DENTAL%20CARIES,dental%20caries%20(tooth%20decay)

Batsford *et al.* (2022) A changing climate and the dental profession. *British Dental Journal.* Volume 232:603–606.
https://www.nature.com/articles/s41415-022-4202-1

Climate change and pregnancy
Yuzen *et al.* (2023) Climate change and pregnancy complications: From hormones to the immune response. *Frontiers Endocrinol.*
https://www.ncbi.nlm.nih.gov/pmc/articles/PMC10113645/pdf/fendo-14-1149284.pdf

Temperature and sperm health
Ai-Phuong Hoang-Thi *et al.* (Apr 2022) The Impact of High Ambient Temperature on Human Sperm Parameters: A Meta-Analysis. *Iran J Public Health.* 51(4): 710–723.
https://www.ncbi.nlm.nih.gov/pmc/articles/PMC9288403/pdf/IJPH-51-710.pdf

Climate and fertility
Sellers, S. & Gray, C. (May 2019) Climate shocks constrain human fertility in Indonesia. *World Dev.* 117;357-369.
https://www.ncbi.nlm.nih.gov/pmc/articles/PMC6581515/

Climate and pregnancy outcomes
Olson, DM. & Metz, GAS. (2020) Climate change is a major stressor causing poor pregnancy outcomes and child development. *F1000Research.*
https://www.ncbi.nlm.nih.gov/pmc/articles/PMC7549179/

Air pollution and adverse birth outcomes
Mitku *et al.* (2023) Impact of ambient air pollution exposure during pregnancy on adverse birth outcomes: generalized structural equation modeling approach. *BMC Public Health* volume 23, Article number: 45.
https://bmcpublichealth.biomedcentral.com/articles/10.1186/s12889-022-14971-3

Rani, Prerna. & Dhok, Archana. (Jan 2023) Effects of Pollution on Pregnancy and Infants. *Cureus.* 15(1): e33906.

https://www.ncbi.nlm.nih.gov/pmc/articles/PMC9937639/#:~:text=Air%20pollution%20may%20indirectly%20harm,with%20organ%20development%20and%20organogenesis.

Climate change and eye disease
Escevarria-Lucas et al. (2012) Impact of climate change on eye diseases and associated economical costs. *In J Environ Res Public Health*. 18:7197.
https://www.ncbi.nlm.nih.gov/pmc/articles/PMC8297364/

Nutrition and the eye
Serhan et al. (Oct 2022) Ophthalmic manifestations of nutritional deficiencies. *J Family Med Prim Care*. 11(10): 5899–5901.
https://www.ncbi.nlm.nih.gov/pmc/articles/PMC9810943/pdf/JFMPC-11-5899.pdf

Climate and retinal detachment
https://europepmc.org/article/med/28549308

The economics of vision loss
Marques et al. (2021) Global economic productivity losses from vision impairment and blindness. *EClinicalMedicine*. 35;100852.
https://www.thelancet.com/action/showPdf?pii=S2589-5370%2821%2900132-2

https://www.cdc.gov/visionhealth/projects/economic_studies.htm

https://bmchealthservres.biomedcentral.com/articles/10.1186/s12913-018-2836-0#:~:text=The%20value%20of%20the%20loss,set%20of%20disability%20weights%20used

Climate change and sleep
Rifkin et al. (2018) Climate change and sleep: A systematic review of the literature and conceptual framework. *Sleep Medicine Reviews* 42;3-9.
https://www.sciencedirect.com/science/article/pii/S1087079218300765?via%-3Dihub

Global sleep problems
Stranges *et al.* (1 Aug 2012) Sleep problems: an emerging global epidemic? Findings from the INDEPTH WHO-SAGE study among more than 40,000 older adults from 8 countries across Africa and Asia. *Sleep.* 35(8):1173-81.
https://www.ncbi.nlm.nih.gov/pmc/articles/PMC3397790/

https://www.ncbi.nlm.nih.gov/books/NBK19961

Insomnia and mental health
Vaughn McCall, W. & Carmen G. Black. (Sept 2013) The Link between Suicide and Insomnia: Theoretical Mechanisms. *Curr Psychiatry Rep.* 15(9): 389.
https://www.ncbi.nlm.nih.gov/pmc/articles/PMC3791319/

Insomnia and thermal regulation
https://sleepeducation.org/sleep-deprivation-disrupts-regulation-body-heat/

Air pollution and insomnia
Liang-Ju Tsai *et al.* (2022) Association between ambient air pollution exposure and insomnia among adults in Taipei City. *Sci Rep.* 12: 19064.
https://www.ncbi.nlm.nih.gov/pmc/articles/PMC9646727/

United Nations, climate change and the risk to children
https://unfccc.int/news/one-billion-children-at-extremely-high-risk-of-the-impacts-of-the-climate-crisis

Air pollution and cancers in young people
https://aacrjournals.org/cebp/article/29/10/1929/124358/Fine-Particulate-Matter-Air-Pollution-and

Heatwaves and the risk to the elderly
https://clinmedjournals.org/articles/jgmg/journal-of-geriatric-medicine-and-gerontology-jgmg-4-053.php

Whitman S *et al*. (Sept 1997) Mortality in Chicago attributed to the July 1995 heat wave. *Am J Public Health*. 87(9): 1515–1518.
https://www.ncbi.nlm.nih.gov/pmc/articles/PMC1380980/

Climate justice
https://www.greenpeace.org.uk/news/climate-change-inequality-climate-justice/

Lenton TM *et al*. (2023) Quantifying the human cost of global warming. Nature Sustainability.
https://www.nature.com/articles/s41893-023-01132-6

Climate change and mental health
Abradovich *et al*. (23 Oct 2018) Empirical evidence of mental health risks posed by climate change *Proc Natl Acad Sci U S A*. 115(43): 10953–10958.
https://www.ncbi.nlm.nih.gov/pmc/articles/PMC6205461/

Peng *et al*. (15 Jul 2017) Effects of ambient temperature on daily hospital admissions for mental disorders in Shanghai, China: A time-series analysis. *Sci Total Environ*. 590-591:281-286.
https://pubmed.ncbi.nlm.nih.gov/28274603/

Global warming and depression
Natur *et al*. (2022) The effect of global warming on complex disorders (mental disorders, primary hypertension and Type 2 diabetes). *Int J Environ Res Public Health*.19:9398.
https://www.ncbi.nlm.nih.gov/pmc/articles/PMC9368177/pdf/ijerph-19-09398.pdf

Climate change and suicide
https://agupubs.onlinelibrary.wiley.com/doi/10.1029/2021GH000580

https://www.imperial.ac.uk/grantham/publications/all-publications/the-impact-of-climate-change-on-mental-health-and-emotional-wellbeing-current-evidence-and-implications-for-policy-and-practice.php

Air pollution and psychiatric illnesses
https://ehp.niehs.nih.gov/doi/full/10.1289/EHP4595

Climate change and substance abuse
Vergunst et al. (2022) Climate Change and Substance-Use Behaviors: A Risk-Pathways Framework. *Perspectives on Psychological Science.* Vol 18(4). https://journals.sagepub.com/doi/full/10.1177/17456916221132739

Solastalgia
https://www.bbc.com/future/article/20151030-have-you-ever-felt-solastalgia

The impact of the alcohol industry on the climate
https://www.journal-of-hepatology.eu/article/S0168-8278(22)00118-0/fulltext#bib32

Air pollution and exercise
Dong Sun et al. (2023) Long term exposure to ambient PM2.5, active commuting, and farm activity and cardiovascular risk in adults in China: A prospective cohort study. *Lancet Planetary Health* 7;e304-12.
https://www.thelancet.com/action/showPdf?pii=S2542-5196%2823%2900047-5

Swimming and heatstroke
https://abcnews.go.com/Health/Wellness/fran-crippen-death-heatstroke-heart-problems/story?id=11967179.

Reduced 24 hour movement and climate change
Zisis et al. (2021) Climate change, 24-hour movement behaviours, and health: a mini umbrella review. *Global Health Research and Policy.* 6;15.

https://ghrp.biomedcentral.com/articles/10.1186/s41256-021-00198-z

Global sport and the climate crisis
https://rapidtransition.org/wp-content/uploads/2020/06/Playing_Against_The_Clock_FINAL.pdf

The pharmaceutical industry and climate change
https://pharmaphorum.com/views-and-analysis/pharmas-climate-change-vulnerability-and-opportunity

Redshaw *et al.* (Jul 2013) Potential changes in disease patterns and pharmaceutical use in response to climate change. *J Toxicol Environ Health B Crit Rev.* 16(5): 285–320.
https://www.ncbi.nlm.nih.gov/pmc/articles/PMC3756629/

Insect vectors and climate change
https://earth.stanford.edu/news/how-does-climate-change-affect-disease#:~:text=As%20the%20globe%20warms%2C%20mosquitoes,chikungunya%20and%20West%20Nile%20virus

Baylis, M. (2017) Potential Impact of climate change on emerging vector-borne and other infections in the UK. *Environmental Health.* 16 suppl;112.
https://pubmed.ncbi.nlm.nih.gov/29219091/

Filho *et al.* (2019) Climate Change, Health and Mosquito-Borne Diseases: Trends and Implications to the Pacific region. *Int J Environ Res Public Health.* 16;5114.
https://www.ncbi.nlm.nih.gov/pmc/articles/PMC6950258/

Harvey *et al.* (2020) Climate change-mediated temperature extremes and insects: From outbreaks to breakdowns. *Glob Change Biol.* 26:6685-6701.
https://onlinelibrary.wiley.com/doi/full/10.1111/gcb.15377

Climate change and antibiotic resistance
https://edition.cnn.com/2023/02/07/health/superbugs-climate-change-scn/index.html#

Air Pollution and antibiotic resistance
Zhou et al. (2023) Association between particulate matter (PM2.5) air pollution and clinical antibiotic resistance: a global analysis. *Lancet Planet Health.* 7;e649-659.
https://www.thelancet.com/action/showPdf?pii=S2542-5196%2823%2900135-3

Antibiotic resistance and raised external temperatures
MacFadden et al. (Jun 2018) Antibiotic Resistance Increases with Local Temperature. *Nat Clim Chang.* 8(6): 510–514.
https://www.ncbi.nlm.nih.gov/pmc/articles/PMC6201249/#:~:text=We%20found%20that%20an%20increase,Klebsiella%20pneumoniae%2C%20and%20Staphylococcus%20aureus.

Climate change and infectious disease risks

Mora et al. (2022) Over half of known human pathogenic diseases can be aggravated by climate change. *Nature Climate Change.* Vol 12;869–875.
https://www.nature.com/articles/s41558-022-01426-1
https://www.cdc.gov/ncezid/what-we-do/climate-change-and-infectious-diseases/index.html

Malaria
https://www.cdc.gov/malaria/about/faqs.html#:~:text=Malaria%20is%20a%20serious%20and,%2C%20and%20flu%2Dlike%20illness.

Increased malaria transmission and temperature
Redshaw et al. (Jul 2013) Potential changes in disease patterns and pharmaceutical use in response to climate change. *J Toxicol Environ Health B Crit Rev.* 16(5): 285–320.

REFERENCES AND NOTES

https://www.ncbi.nlm.nih.gov/pmc/articles/PMC3756629/

Migration of the malaria mosquito
Carlson *et al.* (2023) Rapid range shifts in African *Anopheles* mosquitoes over the last century. *Biology Letters.* 19(2);202203605.
https://royalsocietypublishing.org/doi/epdf/10.1098/rsbl.2022.0365

Lyme disease and climate change
https://climateatlas.ca/lyme-disease-under-climate-change

Leptospirosis and climate change
https://climate-adapt.eea.europa.eu/en/observatory/evidence/health-effects/water-and-food-borne-diseases/leptospirosis-factsheet#:~:text=Higher%20annual%20mean%20temperatures%20enhance,Leptospira%20bacterial%20growth%20and%20survival

Dengue, Chikungunya and Zika
https://www.trexmed.co.uk/dengue-chikungunya-zika/

Dengue fever and climate change
https://www.bmj.com/content/382/bmj.p1690

Dengue mosquitoes in Paris
https://www.theguardian.com/world/2023/sep/01/paris-fumigates-city-tiger-mosquitoes-carry-zika-dengue-disease-france

Cholera
https://www.cdc.gov/cholera/infection-sources.html#:~:text=Global%20Cholera%20Epidemics,-Cholera%20is%20a&text=Each%20year%2C%20an%20estimated%201.3,143%2C000%20die%20from%20it%20worldwide

Cholera, natural disasters and climate change
https://www.pbs.org/newshour/health/as-climate-change-leads-to-more-and-wetter-storms-cholera-cases-are-on-the-rise

Cholera spreading northwards
https://theconversation.com/cholera-cases-are-on-the-rise-and-europe-shouldnt-be-complacent-about-the-risk-203600

Schistosomiasis
https://www.who.int/news-room/fact-sheets/detail/schistosomiasis

Amoebiasis
Shirley, DA. *et al.* (Jul 2018) A Review of the Global Burden, New Diagnostics, and Current Therapeutics for Amebiasis. *Open Forum Infect Dis.* 5(7): ofy161.
https://www.ncbi.nlm.nih.gov/pmc/articles/PMC6055529/

Amoebiasis moving to the USA
https://www.theguardian.com/environment/2022/sep/21/brain-eating-amoeba-climate-crisis-naegleria-fowleri#:~:text=The%20amoeba%20lives%20in%20warm,as%20the%20north%20and%20west

Plague
https://www.cdc.gov/plague/symptoms/index.html

https://archive.ipcc.ch/ipccreports/tar/wg2/index.php?idp=363

Plague and climate change
Tamara Ben Ali. *et al.* (Sept 2011) Plague and Climate: Scales Matter. PLoS Pathog. 7(9): e1002160.
https://www.ncbi.nlm.nih.gov/pmc/articles/PMC3174245/

Chagas disease
https://www.cdc.gov/dpdx/trypanosomiasisamerican/index.html

Chagas disease and temperature
https://archive.ipcc.ch/ipccreports/tar/wg2/index.php?idp=363

REFERENCES AND NOTES

E. Coli and climate change
Philipsbor, R. *et al.* (Jul 2016) Climatic Drivers of Diarrheagenic *Escherichia coli* Incidence: A Systematic Review and Meta-analysis. *J Infect Dis.* 1; 214(1): 6–15.
https://www.ncbi.nlm.nih.gov/pmc/articles/PMC4907410/#:~:text=We%20found%20an%208%25%20increase,increase%20in%20mean%20monthly%20temperature

Cryptosporidiosis and climate change
Wang, XU. *et al.* (2023) Cryptosporidiosis threat under climate change in China: prediction and validation of habitat suitability and outbreak risk for human-derived *Cryptosporidium* based on ecological niche models. *Infectious Diseases of Poverty*. Vol 12, Article number: 35.
https://idpjournal.biomedcentral.com/articles/10.1186/s40249-023-01085-0#:~:text=Under%20future%20climate%20change%2C%20non,%2C%20southwestern%2C%20and%20northwestern%20regions

Giardiasis
https://www.intechopen.com/chapters/57043

Candida and climate change
Ellwanger, JH. & Chies, JAB. (2022) *Candida auris* emergence as a consequence of climate change: Impacts on the Americas and the need to contain greenhouse gas emissions. *The Lancet Regional Health – Americas*. 11:100250.
https://www.thelancet.com/action/showPdf?pii=S2667-193X%2822%2900067-9

Rabies and temperature
Subedi, D. *et al.* (Dec 2022) Ecological and Socioeconomic Factors in the Occurrence of Rabies: A Forgotten Scenario. *Infect Dis Rep.* 14(6): 979–986.
https://www.ncbi.nlm.nih.gov/pmc/articles/PMC9778688/#:~:text=As%20the%20ambient%20temperature%20goes,likely%20to%20spread%20%5B22%5D

Trachoma and climate change
Ramesh A *et al.* (2016) The impact of climate on the abundance of *Musca sorbens*, the vector of trachoma. *Parasites & Vectors* volume 9, Article number: 48.
https://parasitesandvectors.biomedcentral.com/articles/10.1186/s13071-016-1330-y#:~:text=A%20recent%20systematic%20literature%20review,at%20higher%20altitudes%20%5B13%5D

Anthrax and climate warming
Ezhova E *et al.* (June 2021) Climatic Factors Influencing the Anthrax Outbreak of 2016 in Siberia, Russia. *Ecohealth.* 18(2):217-228.
https://pubmed.ncbi.nlm.nih.gov/34453636/

Leprosy in Florida
https://www.bmj.com/content/382/bmj.p1804

The Hummingbird
Editorial. (July 2023) Climate change and health: translating the call of the hummingbird. *The Lancet,* ebioMedicine. Vol 93;104718.
https://www.thelancet.com/action/showPdf?pii=S2352-3964%2823%2900283-9

Important Climate Organisations and Charities

Below is a list of climate change organisations and charities. All are doing their best to warn us of and mitigate against the changes we are making to our climate. I don't think it is possible to be exhaustive and no doubt this list contains my unconscious biases. If there are any I have missed, I apologise. If you have read this book and feel inspired to do something then this list is a good place to start. Browse the various websites and see which organisation (many of whom have local organisations as well as international) best fits your personal ethos.

Climate.gov: https://www.climate.gov
Greenpeace: https://www.greenpeace.org/international/
Friends of the Earth: https://www.foei.org
One Earth: https://www.oneearth.org
World Wildlife Fund (WWF): https://www.oneearth.org
Carbon Brief: https://www.carbonbrief.org/in-depth-qa-what-is-climate-justice/
Mary Robinson Foundation: https://www.mrfcj.org
Asian Pacific Adaption Network: http://www.asiapacificadapt.net
Living Streets: https://www.livingstreets.org.uk

Ummah for Earth: https://ummah4earth.org/en/about-ummah-for-earth/
Climate and Clean Air Coalition: https://www.ccacoalition.org
ClientEarth: https://www.clientearth.org
Oxfam: https://www.oxfam.org.uk/oxfam-in-action/tackling-climate-change/
Rainforest Action Network: https://www.ran.org
Climate Action Network: https://climatenetwork.org
350.org: https://350.org
Climate Alliance: https://www.climatealliance.org/home.html
Oceanic Global: https://oceanic.global
Make My Money Matter: https://makemymoneymatter.co.uk
Union of Concerned Scientists: https://www.ucsusa.org
The Jewish Climate Network: https://www.jcn.org.au
Trees for the Future: https://trees.org
Clean Air Task Force: https://www.catf.us
Christian Climate Action: https://christianclimateaction.org
The Climate Group: https://www.theclimategroup.org
Friends of Nature: http://www.fon.org.cn (Note: this website is in Mandarin)
The European Climate Foundation: https://europeanclimate.org
The Climate Council: https://www.climatecouncil.org.au
Canadian climate Institute/L'Institut Climatique Du Canada: https://climateinstitute.ca
David Suzuki Foundation: https://davidsuzuki.org
New Zealand Climate Action Network: http://www.nzcan.org
Cool Earth: https://www.coolearth.org
The Rainforest Alliance: https://www.rainforest-alliance.org
SeaTrees: https://sea-trees.org
World Forest: https://www.wordforest.org

IMPORTANT CLIMATE ORGANISATIONS AND CHARITIES

Kiko Network (Japan): https://kikonet.org
Wild: https://wild.org

Further Reading, Viewing and Listening

Major Documents and Reports

IPCC Sixth Assessment Report
https://www.ipcc.ch/assessment-report/ar6/

The Royal Society
https://royalsociety.org/topics-policy/energy-environment-climate/

UN Environment Programme Annual Report
https://www.unep.org/resources/annual-report-2022?gad_source=1&gclid=EAIaIQobChMIrpe_7sqYgwMVbpdQBh0My-QC7EAAYASAAEgL9r_D_BwE

The Stern Review
https://webarchive.nationalarchives.gov.uk/ukgwa/20100407172811/https:/www.hm-treasury.gov.uk/stern_review_report.htm

World Health Organization
https://www.who.int/news-room/fact-sheets/detail/climate-change-and-health

World Meteorological Organization – State of the Global Climate
https://www.who.int/news-room/fact-sheets/detail/climate-change-and-health

Non-fiction Books

Lenton, T., Watson, A. (2011) *Revolutions That Made the Earth.* Oxford University Press.
Gates, B. (2021) *How to Avoid a Climate Disaster: The Solutions We Have and the Breakthroughs We Need.* Penguin Publishers.
Wallace-Wells, D. (2019) *The Uninhabitable Earth: A Story of the Future.* Penguin Publishers.
Thunberg, G. (2022) *The Climate Book.* Penguin Publishers.
Lovelock, J. (2015) *A Rough Guide to the Future.* Penguin Science.
Lynas, M. (2020) *Our Final Warning. Six Degrees of Climate Emergency.* Fourth Estate.
Berners-Lee, M. (2021) *There is No Planet B.* Cambridge University Press.

Climate Fiction

Stanley Robinson, K. (2020) *The Ministry for the Future.* Orbit.
Ballard, JG (2009) *The Drought.* Fourth Estate.
Ballard, JG. (2010) *The Drowned World.* Fourth Estate.
Le Guin, U. (1989) *The New Atlantis.* Tor Books.
Adams, J. (2015) *Loosed Upon the World.* Saga Press.
Powers, R. (2022) *The Overstory.* Vintage Classics.

Films and TV

Don't Look Up. (2021) Netflix.
Wall-E. (2008) Disney/Pixar.
The Day after Tomorrow. (2004) 20th Century Fox.
An Inconvenient Truth. (2006) Al Gore, Various platforms.
A Life on Our Planet. (2020) David Attenborough, Netflix.
Burning, Eva Orner. (2021) Amazon Prime.

Podcasts

TILclimate, Apple and Spotify.
For What It's Earth, Apple and Spotify.
America Adapts, Apple and Spotify.
The Climate Question, Apple, Spotify, and BBC Sounds.
How to Save a Planet, Apple and Spotify.
Outrage and Optimism, Apple and Spotify.

Acknowledgements

I would like to thank all those at Black Spring Press Group (BSPG) for their invaluable help with this book. In particular, I would like to thank Todd Swift for believing both in the importance and urgency of its message.